Cambridge O Level

Chemistry

Bryan Earl
Doug Wilford

Boost

HODDER
EDUCATION
AN HACHETTE UK COMPANY

The Publishers would like to thank the following for permission to reproduce copyright material.

Acknowledgements

Cambridge International copyright material in this publication is reproduced under licence and remains the intellectual property of Cambridge Assessment International Education.

Cambridge Assessment International Education bears no responsibility for the example answers to questions taken from its past question papers which are contained in this publication.

Exam-style questions (and sample answers) have been written by the authors. In examinations, the way marks are awarded may be different. References to assessment and/or assessment preparation are the publisher's interpretation of the syllabus requirements and may not fully reflect the approach of Cambridge Assessment International Education.

Third-party websites and resources referred to in this publication have not been endorsed by Cambridge Assessment International Education.

Every effort has been made to trace all copyright holders, but if any have been inadvertently overlooked, the Publishers will be pleased to make the necessary arrangements at the first opportunity.

Although every effort has been made to ensure that website addresses are correct at time of going to press, Hodder Education cannot be held responsible for the content of any website mentioned in this book. It is sometimes possible to find a relocated web page by typing in the address of the home page for a website in the URL window of your browser.

We have carried out a health and safety check of this text and have attempted to identify all recognised hazards and suggest appropriate cautions. However, the Publishers and the authors accept no legal responsibility on any issue arising from this check; whilst every effort has been made to carefully check the instructions for practical work described in this book, it is still the duty and legal obligation of schools to carry out their own risk assessments for each practical in accordance with local health and safety requirements.

For further health and safety information (e.g. Hazcards) please refer to CLEAPSS at www.cleapss.org.uk.

Hachette UK's policy is to use papers that are natural, renewable and recyclable products and made from wood grown in well-managed forests and other controlled sources. The logging and manufacturing processes are expected to conform to the environmental regulations of the country of origin.

Orders: please contact Hachette UK Distribution, Hely Hutchinson Centre, Milton Road, Didcot, Oxfordshire, OX11 7HH. Telephone: +44 (0)1235 827827. Email education@hachette.co.uk. Lines are open from 9 a.m. to 5 p.m., Monday to Friday. You can also order through our website: www.hoddereducation.co.uk

ISBN: 9781398310599

© Bryan Earl and Doug Wilford 2021

First published in 2021 by
Hodder Education
An Hachette UK Company
Carmelite House
50 Victoria Embankment
London EC4Y 0DZ

www.hoddereducation.com

Impression number   10 9 8 7 6 5 4 3 2 1

Year   2025 2024 2023 2022 2021

Cover photo © Björn Wylezich - stock.adobe.com

Illustrations by Integra Software Services Pvt. Ltd., Pondicherry, India

Typeset in Integra Software Services Pvt. Ltd., Pondicherry, India

Printed in Great Britain by Bell and Bain Ltd, Glasgow

A catalogue record for this title is available from the British Library.

Contents

Acknowledgements

The authors would like to thank Irene, Katharine, Michael and Barbara for their patience, support and encouragement throughout the production of this textbook. We would also like to thank the editorial and publishing teams at Hodder Education who have supported us on the journey over the past year.

Source acknowledgements

pp.13, 37, 39, 40, 41, 183, 188, 189, 192, 193 and 206. The molecular models shown were made using the Molymod® system available from Molymod® Molecular Models, Spiring Enterprises Limited, Billingshurst, West Sussex RH14 9NF England.

Photo credits

r right, *l* left, *t* top, *m* middle, *b* bottom
p.1 © *bl* Dolphinartin/stock.adobe.com, *tr* © 12ee12/stock.adobe.com, *br* © Olaf Schubert/Image Broker/Alamy Stock Photo; **p.2** *l* © Donald L Fackler Jr/Alamy Stock Photo, *r* © Andrew Lambert Photography/Science Photo Library; **p.3** *tl* © Geoscience Features Picture Library/Dr.B.Booth, *bl* © Javier Larrea/Agefotostock/Alamy Stock Photo, *br* © Power And Syred/Science Photo Library; **p.7** *l* © Andrew Lambert Photography/Science Photo Library; **p.8** *l* © Giphotostock/Science Photo Library, *r* © Giphotostock/Science Photo Library; **p.10** © NASA/ESA/STSCI/Hubble Heritage Team/Science Photo Library; **p.11** *t* © Robert Harding/Getty Images, *m* © FlemishDreams – Fotolia, *b* © Echo 23 Media/Alamy Stock Photo; **p.12** *tl* © Bay Ismoyo/AFP/Getty Images, *bl* © Christie's Images / The Bridgeman Art Library, *r* © BL Images Ltd / Brian Lawrence / Alamy Stock Photo; **p.13** © Andrew Lambert Photography/Science Photo Library; **p.15** © Andrew Lambert Photography / Science Photo Library; **p.18** *l* © Galyna Andrushko – Fotolia, *br* © Andrew Lambert Photography / Science Photo Library, *tr* © Andrew Lambert Photography / Science Photo Library; **p.19** *l-r* © Vvvita/stock.adobe.com, © Weyo/stock.adobe.com, © Shutterstock/Stay_Positive, © Magraphics/stock.adobe.com, © sommai/stock.adobe.com ; **p.20** *tr* © Science Photo Library/Eye of Science, *bl* © moodboard/Thinkstock, *tr* © Digital Instruments/Veeco/Science Photo Library; **p.24** © Henry Westheim Photography / Alamy; **p.30** *tl* © Andrew Lambert Photography/ Science Photo Library, *tr* © Andrew Lambert Photography/ Science Photo Library, *b* © The_lightwriter/stock.adobe.com; **p.35** *l* ©

Martyn F. Chillmaid, *r* © Martyn F. Chillmaid; **p.36** *(left column)* *l* © Martyn F. Chillmaid, *r* © Martyn F. Chillmaid, *(right column)* Courtesy of the University of Illinois at Urbana-Champaign Archives; **p.38** © Andrew Lambert Photography/Science Photo Library; **p.39** © Andrew Lambert Photography/Science Photo Library; **p.40** *l* © Andrew Lambert Photography/ Science Photo Library, *r* © Andrew Lambert Photography/Science Photo Library; **p.41** © Andrew Lambert Photography/Science Photo Library; **p.43** © E.R.Degginger/Science Photo Library; **p.45** *l* © Philippe Plailly/Science Photo Library, *r* © Paul Davey/Alamy Stock Photo; **p.47** *tl* © Richard Megna/Fundamental/Science Photo Library, *bl* © Sheila Terry/Science Photo Library, *tr* © ACORN 1/Alamy Stock Photo, *br* © Kage Mikrofotografie GBR/Science Photo Library; **p.48** © overthehill – Fotolia; **p.54** *l* © Andrew Lambert Photography/Science Photo Library, *r* © Andrew Lambert Photography/Science Photo Library; **p.55** © Andrew Lambert Photography/ Science Photo Library; **p.61** © Mazz Pro/ Shutterstock.com; **p.66** *tl* © Alexander Maksimenko – Fotolia, *tr* © Mkos83/stock.adobe.com, *b* © Pavel Losevsky/stock.adobe.com; **p.69** © Howard Davies/Alamy Stock Photo; **p.70** © Kekyalyaynen/Shutterstock.com; **p.71** *l* © dpa picture alliance archive / Alamy Stock Photo, *r* © Opsorman/Shutterstock.com; **p.72** *l* © Kathy Gould/iStock/Thinkstock, *r* © Trevor Clifford Photography/Science Photo Library; **p.75** *t* © Andrew Lambert Photography/Science Photo Library, *b* © Clynt Garnham Renewable Energy / Alamy Stock Photo; **p.78** © Gchutka/E+/Getty Images; **p.79** © Adam Hart-Davis/Science Photo Library; **p.80** *t* © Andrew Lambert Photography/Science Photo Library, *b* © Andrew Lambert Photography/Science Photo Library; **p.83** *tl* © Dinodia Photos / Alamy Stock Photo, *bl* © Oleg Zhukov – Fotolia, *tm* © Matt K – Fotolia, *mb* © Dmitry Sitin – Fotolia, *tr* © iofoto – Fotolia, *br* © yang yu – Fotolia; **p.84** *bl* © Andrew Lambert Photography/Science Photo Library, *tr* © Hkhtt hj/Shutterstock.com; **p.85** © Stevie Grand/Science Photo Library; **p.86** © GeoScience Features Picture Library; **p.97** *tl* © OutdoorPhotos - Fotolia, *bl* © Arnaud SantiniI – Fotolia, *tr* © Yakovlevadaria/stock.adobe.com, *br* © Motoring Picture Library / Alamy; **p.98** *l* © Andrew Lambert Photography/Science Photo Library, *tr* © Andrew Lambert Photography/ Science Photo Library, *br* © Andrew Lambert Photography/Science Photo Library; **p.100** *tr* ©

Chillmaid. Thanks to Molymod.com for providing the model, *bl* © Andrew Lambert Photography/Science Photo Library, *tr* © Steve Cukrov – Fotolia, *br* © Andrew Lambert Photography/Science Photo Library; **p.194** © smikeymikey1 – Fotolia; **p.195** *tl* © Henning Kaiser/ DPA/PA Images, *bl* © Richard Carey/ stock.adobe.com, *tr* © Sally Morgan/Ecoscene, *br* © Andrew Lambert Photography/Science Photo Library; **p.199** © M.studio – Fotolia; **p.201** © Ian Dagnall/ Alamy Stock Photo; **p.202** *t* © Biophoto Associates/ Science Photo Library, *b* © Science Photo Library; **p.203** *l* © Andrew Lambert Photography/Science Photo Library, *r* © Andrew Lambert Photography/ Science Photo Library; **p.206** *tl–bl (table)* © Martyn F Chillmaid. Thanks to Molymod.com for providing the models, *b* © Paul Cooper / Rex Features; **p.208** *l* © Andrew Lambert Photography/Science Photo Library, *r* © Leonid Shcheglov – Fotolia; **p.212** © burnel11 – Fotolia; **p.213** *tl* © Martyn F Chillmaid, *tr* © Martyn F Chillmaid, *br* © Martyn F Chillmaid; **p.214** © Maurice Savage/Alamy Stock Photo; **p.217** *bl* © Vectorfusionart/stock.adobe.com, *tr* © Klaus Guldbrandsen/Science Photo Library, *br* © Klaus Guldbrandsen/Science Photo Library; **p.218** *l* © Andrew Lambert Photography/Science Photo Library, *tr* © Ricardo Funari / Brazilphotos/Alamy Stock Photo, *br* © Andrew Lambert Photography/Science Photo Library; **p.219** *l* © Sipa Press / Rex Features, *tr* © KPA/Zuma/Rex Features, *br* © Sciencephotos/ Alamy Stock Photo; **p.220** © BOC Gases; **p.221** *tl* © Paul Rapson/Science Photo Library, *bl* © Jenny Matthews/Alamy Stock Photo, *tr* © Andrew Lambert Photography/Science Photo Library, **p.222** © Andrew Lambert Library; **p.224** © Niall McDiarmid / Alamy; **p.225** © Ggw/stock.adobe.com; **p.226** *(left column) l* © Turtle Rock Scientific/Science Source/Science Photo Library *r* © Science Photo Library, *(right column) r* © Giphotostock/Science Photo Library; **p.227** *l* © Andrew Lambert Photography/Science Photo Library, *r* © Martyn F. Chillmaid/Science Photo Library.

How to use this book

To make your study of Chemistry for Cambridge O Level as rewarding and successful as possible, this textbook, endorsed by Cambridge Assessment International Education, offers the following important features:

FOCUS POINTS

Each topic starts with a bullet point summary of what you will encounter within the topic.

This is followed by a short outline of the topic so that you know what to expect over the next few pages.

Test yourself

These questions appear regularly throughout the chapter so you can check your understanding as you progress.

Revision checklist

At the end of each chapter, a revision checklist will allow you to recap what you have learned in each topic and double-check that you understand the key concepts before moving on.

Exam-style questions

Each chapter is followed by exam-style questions to help familiarise learners with the style of questions they may see in their examinations. These will also prove useful in consolidating your learning. Past paper questions are also provided at the end of the book.

Key definitions

These provide explanations of the meanings of key words as required by the syllabus.

Practical skills

These boxes identify the key practical skills you need to understand as part of completing the course.

? Worked examples

These boxes give step-by-step guidance on how to approach different sorts of calculations, with follow-up questions so you can practise these skills.

Going further

These boxes take your learning further than is required by the Cambridge syllabus so that you have the opportunity to stretch yourself.

Answers are provided online at www.hoddereducation.com/cambridgeextras

Scientific enquiry

Throughout your O Level Chemistry course you will need to carry out experiments and investigations aimed at developing some of the **skills** and **abilities** that scientists use to solve real-life problems.

Simple experiments may be designed to measure, for example, the temperature of a solution or the rate of a chemical reaction. Longer experiments, or investigations, may be designed to allow you to actually see the relationship between two or more physical quantities such as how rate of reaction varies with temperature and concentration.

Investigations are likely to come about from the topic you are currently studying in class, and your teacher may provide you with suggestions. For all investigations both your teacher and you must consider the safety aspects of the chemicals and apparatus involved. You should never simply carry out a chemistry investigation without consideration of the hazards of the chemicals or getting the approval of your teacher.

To carry out an investigation you will need to:

1 **Select and safely use suitable techniques, apparatus and materials** – your aim must be to safely collect sufficient evidence using the most appropriate apparatus for the technique you have chosen. Being able to draw and label diagrams correctly to show how the equipment will be used is also important. Your techniques will need to be explained clearly to do a proper risk assessment. For example, how to carry out a titration, how you are going to follow the rate of the reaction you are using, or how to test for ions and gases.

2 **Plan your experiment** – this is an important part of doing science and involves working out what you are going to do to try to find answers to the questions you have set yourself. Predictions based on work you have been studying or are doing in class may help you develop the investigation in terms of the number and type of observations or data needed. You will also need to be able to identify the independent and dependent variables. For example, if you are trying to find out how temperature affects the rate of a reaction, the temperature will be the independent variable, but the dependent variable might be the volume of gas collected. Other variables such as concentration need to be controlled so that they will not affect the data obtained. Most importantly, it is essential that you carry out a risk assessment before you do any practical work.

3 **Make and record observations** – the data you need to answer the questions you have set yourself can only be found if you have planned your investigation sensibly and carefully. For example, you might start to use a measuring cylinder to collect a gas, but as you develop your ideas you may realise a burette might be more appropriate and more accurate. Be careful not to dismantle the equipment/apparatus until you have completed your analysis of the data, and you are sure you do not need to repeat any of the measurements! If you have to reset your equipment/apparatus it may add further errors to the results you have obtained.

Ensure that all your data, numeric or observational, is displayed in a clear format. This will often be in the form of headed tables with the correct units being shown to the appropriate degree of precision.

4 **Interpret and evaluate observations** – the results you obtain from any investigation must be displayed carefully and to the accuracy of the equipment you have chosen to use. Your choice of presentation will help you interpret your evidence and make conclusions. Often your presentation will be in the form of a graph or a table. For some graphs you may need to calculate gradients or use it to find values at a specific point during the investigation by drawing intercepts. Good chemists keep looking at the data and alter the way in which it is obtained to get more accurate results. You should be able to evaluate whether your data is good or bad. If it is good, were there any anomalous results? Why did you get them?

5 **Evaluate methods and suggest possible improvements** – at the end of your investigation you must be able to evaluate the equipment, methods and techniques that you have used. Think about any sources of errors that could have affected your results by the use of the wrong equipment. Consider, if you were able to carry out the investigation again, what you would change. The more data you obtain the easier it is to spot anomalous results.

A **written report** of any chemical investigation would normally be made of these fixed components:

» First, state the **aim** of the work at the very beginning to inform your teacher what you were doing the investigation to find out.

» A list of all items of equipment/apparatus used and a record of the smallest division of the scale of each measuring device you have used (see Chapter 14). For example, burettes can be read to two decimal places, to the nearest $0.05\,cm^3$, where the second decimal place is either a 0 if the bottom of the meniscus is on the scale division, or a 5 if it is between the divisions. If the meniscus was between $24.10\,cm^3$ and $24.20\,cm^3$ the reading would be $24.15\,cm^3$.

» You must show that you have considered the safety of yourself and others before you carry out any practical work. Provide a list of all the chemicals you will use, as well as the ones you will produce, and do a **risk assessment** to check on all the hazards of the chemicals. The results of your risk assessment might indicate that you need to work in a well-ventilated room or in a fume cupboard. In some cases, you may need the assistance of your teacher. If in doubt, always ask your teacher for advice.

» Clearly state the details of the **methods** used, starting with the wearing of eye protection. The methods should be shown as numbered steps and should be made as clear as possible. Ideas of the number of measurements that will be made and their frequency should be stated. Observations should be clear and you should use changes in colour and physical state as part of your observations.

» Presentation of **results** and **calculations**. If you made several measurements of a quantity, draw up a table in which to record your results. Use the column headings, or start of rows, to name exactly what the measurement is and state the units used; for example in a rates of reaction experiment, 'Mass of calcium carbonate/g'.

Give numeric values to the number of significant figures appropriate to the equipment being used, for example a mass could be recorded to 0.5 g or 0.05 g depending on the resolution of the top-pan balance you use. Take averages and remember that anomalous or non-concordant results should not be used in their calculation. If you decide to make a graph of your results you will need at least six data points taken over as large a range as possible; be sure to label

each axis of a graph with the name and unit of the quantity being plotted. Make sure that the scale you use allows the points to fill up as much of the graph paper as possible.

Clearly explain the calculations involved in the interpretation of your data and give the significant figures appropriate to the equipment used.

» **Conclusions** can be obtained from the graphs and calculations you carry out. Your conclusions from the data obtained might be different from those that you expected. Even so it is very important for any scientist to come to terms with the findings of their experimental result, good or bad!

» In the **evaluation** you should make a comparison between the conclusions of your investigation and your expectations: how close or how different were they? You should comment on the reliability and accuracy of the observations and the data obtained. Could you have improved the method to give better or more accurate results? Would a pH probe have been better than using universal indicator to find the point of neutralisation in a titration? Were there any anomalous points on your graphs, or any unusual data or observations? Highlight these and try to give an explanation.

Suggestions for investigations

Some suggested investigations are outlined in this book as follows:

1 Find which vinegars contain the most acid. (Chapter 8)
2 Find the molar volume of hydrogen by reacting magnesium with hydrochloric acid. (Chapter 4)
3 Determination of the enthalpy of combustion of ethanol. (Chapter 6)
4 The effect of changing the surface area of a reactant on the rate of reaction. (Chapter 7)
5 Use the anion and cation methods of identification to find the ions present in tap water. (Chapter 14)
6 Show that ammonia is a weak base by measuring its pH and conductivity, and comparing your results with those from a solution of sodium hydroxide with the same concentration. (Chapter 8)
7 Determine the melting point of stearic acid. (Chapter 1)

8 How can sodium chloride be obtained from rock salt? (Chapter 14)

9 What are the effects of acid rain on a variety of building materials? (Chapter 11)

10 Which is the best temperature, between 34 and 40°C, for the fermentation of sugar to take place? (Chapter 12)

11 What are the chemical properties of the weak organic acid, ethanoic acid? (Chapter 13)

12 Do foodstuffs contain carbon? (Chapter 12)

Ideas and evidence in science

In some of the investigations you perform in the school laboratory, you may find that you do not interpret your data in the same way as your friends do; perhaps you will argue with them as to the best way to explain your results and try to convince them that your interpretation is right. Scientific controversy frequently arises through people interpreting evidence differently.

For example, our ideas about atoms have changed over time. Scientists have developed new models of atoms over the centuries as they collected new experimental evidence. If we go back to the Greeks in the 5th century BC, they thought matter was composed of indivisible building blocks which they called *atomos*. However, the idea was essentially forgotten for more than 2000 years. Then John Dalton published his ideas about atoms in 1800. He suggested that all matter was made of tiny particles called atoms, which he imagined as tiny spheres that could not be divided. It then took another 100 years before Joseph Thomson, Ernest Rutherford and James Chadwick carried out experiments and discovered that there was a structure within the atom. This saw the continuous development of what we know today as atomic theory.

A further example involves the well-known Russian chemist Dimitri Mendeleev. He realised that the physical and chemical properties of the known elements were related to their atomic mass in a 'periodic' way, and arranged them so that groups of elements with similar properties fell into vertical columns in his table. However, in devising his table, Mendeleev did not conform completely to the order of atomic mass, with some elements swapped around. It took time for his ideas to gain acceptance because the increase in atomic mass was not regular when moving from one element to another. We now know, with the development of atomic theory and a better understanding of chemical processes, that the elements in the Periodic Table are not all in atomic mass order. It took until 1934, with an understanding of atomic number and post-Russian revolution, for the Periodic Table to be finally accepted in the form you see today.

There are many different types of scientists with specialties in their own areas of work such as chemistry and physics, but they all work in the same way. They come up with new theories and ideas, they carry out work to find the evidence to establish whether their ideas are correct and, if not, why. Scientists rely on other scientists checking their work, often improving the ideas of everyone and moving science forward. The use of new ideas is often beneficial to everyone in the world, for example the discovery of vaccines for Covid-19, or the push to improve battery manufacture for use in electric cars which would in turn help solve one of the biggest problems we have to face: global warming. Scientists are working hard to stop global warming but their ideas are not always embraced because of economic and political factors.

1 States of matter

FOCUS POINTS
★ What is the structure of matter?
★ What are the three states of matter?
★ How does kinetic particle theory help us understand how matter behaves?

In this first chapter you will look at the three states of matter: solids, liquids and gases. The structure of these states of matter and how the structures can be changed from one to another is key to understanding the states of matter.

You will use the kinetic particle theory to help explain how matter behaves, so you can understand the difference in the properties of the three states of matter and how the properties are linked to the strength of bonds between the particles they contain. Why, for example, can you compress gases but cannot compress a solid? By the end of this chapter you should be able to answer this question, and use the ideas involved to help you to understand many everyday observations, such as why car windows mist up on a cold morning or why dew forms on grass at night.

1.1 Solids, liquids and gases

Chemistry is about what **matter** is like and how it behaves, and our explanations and predictions of its behaviour. What is matter? This word is used to cover all the substances and materials from which the physical universe is composed. There are many millions of different substances known, and all of them can be categorised as solids, liquids or gases (Figure 1.1). These are what we call the three **states of matter**.

b Liquid

a Solid

c Gas

▲ **Figure 1.1** Water in three different states

A solid, at a given temperature, has a definite volume and shape which may be affected by changes in temperature. Solids usually increase slightly in size when heated, called expansion (Figure 1.2), and usually decrease in size if cooled, called contraction.

A liquid, at a given temperature, has a fixed volume and will take the shape of any container into which it is poured. Like a solid, a liquid's volume is slightly affected by changes in temperature.

A gas, at a given temperature, has neither a definite shape nor a definite volume. It will take the shape of any container into which it is placed and will spread evenly within it. Unlike solids and liquids, the volumes of gases are affected greatly by changes in temperature.

Liquids and gases, unlike solids, are compressible. This means that their volume can be reduced by the application of pressure. Gases are much more compressible than liquids.

▲ **Figure 1.2** Without expansion gaps between the rails, the track would bend when it expanded in hot weather

1.2 The kinetic particle theory of matter

The **kinetic particle theory** helps to explain the way that matter behaves. It is based on the idea that all matter is composed of tiny particles. This theory explains the physical properties of matter in terms of the movement of the particles from which it is made.

The main points of the theory are:

» All matter is composed of tiny, moving particles, invisible to your eye. Different substances have different types of particles (atoms, molecules or ions) of varying sizes.
» The particles move all the time. The higher the temperature, the faster they move on average.
» Heavier particles move more slowly than lighter ones at a given temperature.

The kinetic particle theory can be used as a scientific model to explain how the arrangement of particles relates to the properties of the three states of matter.

Explaining the states of matter

In a solid the particles attract one another. There are attractive forces between the particles which hold them close together. The particles have little freedom of movement and can only vibrate about a fixed position. They are arranged in a regular manner, which explains why many solids form **crystals**.

It is possible to model such crystals by using spheres to represent the particles. For example, Figure 1.3a shows spheres built in a regular way to represent the structure of a chrome alum crystal. The shape is very similar to that of a part of an actual chrome alum crystal (Figure 1.3b).

a   A model of a chrome alum crystal

▲ **Figure 1.5** Sodium chloride crystals

In a liquid, the particles are still close together but they move in a random way and often collide with one another. The forces of attraction between the particles in a liquid are weaker than those in a solid. Particles in the liquid form of a substance have more energy on average than the particles in the solid form of the same substance.

In a gas, the particles are relatively far apart. They are free to move anywhere within the container in which they are held. They move randomly at very high velocities, much more rapidly than those in a liquid. They collide with each other, but less often than in a liquid, and they also collide with the walls of the container. They exert virtually no forces of attraction on each other because they are relatively far apart. Such forces, however, are very significant. If they did not exist, we could not have solids or liquids (see Changes of state, p. 4).

b An actual chrome alum crystal

▲ **Figure 1.3**

Studies using X-ray crystallography (Figure 1.4) have confirmed how particles are arranged in crystal structures. When crystals of a pure substance form under a given set of conditions, the particles are always arranged (or packed) in the same way. However, the particles may be packed in different ways in crystals of different substances. For example, common salt (sodium chloride) has its particles arranged to give cubic crystals as shown in Figure 1.5.

The arrangement of particles in solids, liquids and gases is shown in Figure 1.6.

Solid
Particles only vibrate about fixed positions. Regular structure.

Liquid
Particles have some freedom and can move around each other. Collide often.

Gas
Particles move freely and at random in all the space available. Collide less often than in liquid.

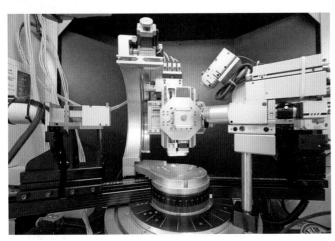

▲ **Figure 1.4** A modern X-ray crystallography instrument used for studying crystal structure

▲ **Figure 1.6** The arrangement of particles in solids, liquids and gases

1.3 Changes of state

The kinetic particle theory model can be used to explain how a substance changes from one state to another. If a solid is heated, the particles vibrate faster as they gain energy. This makes them 'push' their neighbouring particles further away. This causes an increase in the volume of the solid, such that the solid expands, and we can say that expansion has taken place.

Eventually, the heat energy causes the forces of attraction to weaken. The regular pattern of the structure breaks down, and the particles can now move around each other. The solid has melted. The temperature at which this takes place is called the **melting point** of the substance. The temperature of a melting pure solid will not rise until it has all melted. When the substance has become a liquid there are still very significant forces of attraction between the particles, which is why the substance is a liquid and not a gas.

Solids which have high melting points have stronger forces of attraction between their particles than those which have low melting points. A list of some substances with their corresponding melting and boiling points is shown in Table 1.1.

▼ **Table 1.1** Melting points and boiling points of substances

Substance	Melting point/°C	Boiling point/°C
Aluminium	661	2467
Ethanol	−117	79
Magnesium oxide	827	3627
Mercury	−30	357
Methane	−182	−164
Oxygen	−218	−183
Sodium chloride	801	1413
Sulfur	113	445
Water	0	100

If a liquid is heated, the average energy of the particles increases and the particles will move even faster. Some particles at the surface of the liquid have enough energy to overcome the forces of attraction between themselves and the other particles in the liquid and they escape to form a gas. The liquid begins to **evaporate** as a gas is formed.

Eventually, a temperature is reached at which the particles are trying to escape from the liquid so quickly that bubbles of gas actually start to form inside the liquid. This temperature is called the **boiling point** of the substance. At the boiling point, the pressure of the gas created above the liquid equals that of the air, which is **atmospheric pressure**.

Liquids with high boiling points have stronger forces between their particles than liquids with low boiling points.

When a gas is cooled, the average energy of the particles decreases and the particles move closer together. The forces of attraction between the particles now become significant and cause the gas to **condense** into a liquid. When a liquid is cooled, it freezes to form a solid. Energy is released in each of these changes.

Changes of state are examples of **physical changes**. Whenever a physical change of state occurs, the temperature remains constant during the change. During a physical change, no new substance is formed.

Heating and cooling curves

The graph shown in Figure 1.7 was drawn by plotting the temperature of water as it was heated steadily from −15°C to 110°C. You can see from the curve that changes of state have taken place. When the temperature was first measured, only ice was present. After a short time, the curve flattens showing that even though heat energy is being put in, the temperature remains constant.

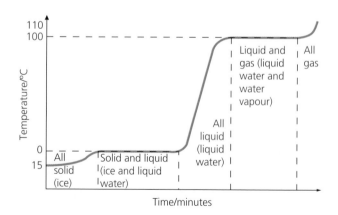

▲ **Figure 1.7** Graph of temperature against time for the change from ice at −15°C to water to steam

Practical skills

Changes of state

Safety

- Eye protection must be worn.
- Take care when handling and using hot water.

The apparatus below was set up to obtain a cooling curve for stearic acid. The stearic acid was placed into a boiling tube which was then placed in a beaker of water that was heated to 80°C, which is above the melting point of stearic acid.

The boiling tube was then removed from the beaker and the temperature of the stearic acid was recorded every minute for 12 minutes (see table below) using the thermometer to stir the stearic acid while it was a liquid.

1 Why was it important to remove the boiling tube with the stearic acid from the water?
2 Why was the stearic acid stirred with the thermometer?
3 Why were temperature readings taken every minute for 12 minutes?
4 Draw and label axes for plotting this data.
5 Plot the points and draw a line of best fit.
6 a At what temperature did the stearic acid begin to change state?
   b How could you tell this from your graph?
   c Explain what is happening at this temperature.

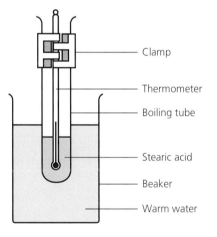

Clamp

Thermometer

Boiling tube

Stearic acid

Beaker

Warm water

Time/mins	0	1	2	3	4	5	6	7	8	9	10	11	12
Temperature/°C	79	76	73	70	69	69	69	69	69	67	64	62	60

In ice, the particles of water are close together and are attracted to one another. For ice to melt, the particles must obtain sufficient energy to overcome the forces of attraction between the water particles, so that relative movement can take place. The heat energy is being used to overcome these forces.

The temperature will begin to rise again only after all the ice has melted. Generally, the heating curve for a pure solid always stops rising at its melting point and produce a sharp melting point.

A sharp melting point therefore indicates that it is a pure sample. The addition or presence of impurities lowers the melting point.

You can find the melting point of a substance using the apparatus shown in Figure 1.8. The addition or presence of impurities lowers the melting point. A mixture of substances also has a lower melting point than a pure substance, and the melting point will be over a range of temperatures and not sharp.

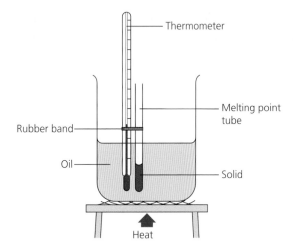

▲ **Figure 1.8** If a substance, such as the solid in the melting point tube, is heated slowly, this apparatus can be used to find the melting point of the substance

In the same way, if you want to boil a liquid, such as water, you have to give it some extra energy. This can be seen on the graph in Figure 1.7, where the curve levels out at 100°C – the boiling point of water.

Solids and liquids can be identified from their characteristic melting and boiling points.

The reverse processes of condensing and freezing occur when a substance is cooled. Energy is given out when the gas condenses to the liquid and the liquid freezes to give the solid.

1.4 The effects of temperature and pressure on the volume of a gas

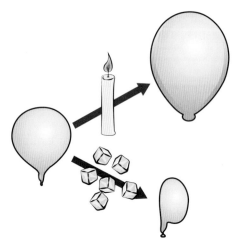

▲ **Figure 1.9** Temperature changes the volume of the air in a balloon. Higher temperatures increase the volume of the balloon and cold temperatures reduce its volume.

What do you think has caused the difference between the balloons in Figure 1.9? The pressure inside a balloon is caused by the gas particles striking the inside surface of the balloon (Figure 1.10). At a higher temperature there is an increased pressure inside the balloon. This is due to the gas particles having more energy and therefore moving around faster, which results in the particles striking the inside surface of the balloon more frequently, which leads to an increase in pressure.

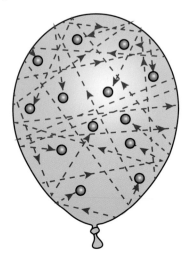

▲ **Figure 1.10** The gas particles striking the surface create the pressure

Since the balloon is made from an elastic material, the increased pressure causes the balloon to stretch and the volume increases. An increase in volume of a gas with increased temperature is a property of all gases. French scientist J.A.C. Charles made an observation like this in 1781 and concluded that when the temperature of a gas increased, the volume also increased at a fixed pressure. We can extend this idea to suggest that changing the pressure of a fixed volume of a gas must have an effect on the temperature of the gas. If you have ever used a bicycle pump to blow up a bicycle tyre then you may have felt the pump get hotter the more you used it. As you use the pump you increase pressure on the air in the pump. Such an increase in pressure causes the gas molecules to move closer together and the molecules to collide more frequently. As a result, more frictional forces come into play, and this causes the temperature to rise. In addition, as the molecules are forced closer to one another, intermolecular bonds form, again increasing the temperature of the gas. As the temperature of the gas increases, this also causes the molecules to move faster, causing even more collisions.

Test yourself

2 Why do gases expand more than solids for the same increase in temperature?
3 Ice on a car windscreen will disappear as you drive, even without the heater on. Explain why this happens.
4 When salt is placed on ice, the ice melts. Explain why this happens.
5 Draw and label a graph of water at 100°C being allowed to cool to –5°C.

1.5 Diffusion

When you go through the door of a restaurant you can often smell the food being cooked. For this to happen, gas particles must be leaving the pans the food is being cooked in and be spreading through the air in the restaurant. This spreading of a gas is called **diffusion** and it occurs in a haphazard and random way.

All gases diffuse to fill the space available. Figure 1.11 shows two gas jars on top of each other. Liquid bromine has been placed in the bottom gas jar (left photo) and then left for a day (right photo). The brown-red fumes are gaseous bromine that has spread evenly throughout both the gas jars from the liquid present in the lower gas jar.

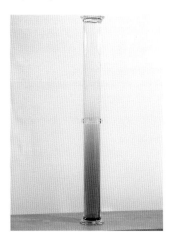

▲ **Figure 1.11** After 24 hours the bromine fumes have diffused throughout both gas jars

Diffusion can be explained by the **kinetic particle theory**. This theory states that all matter is made of many small particles which are constantly moving. In a solid, as we have seen, the particles simply vibrate about a fixed point. However, in a gas, the particles move randomly past one another, colliding with each other.

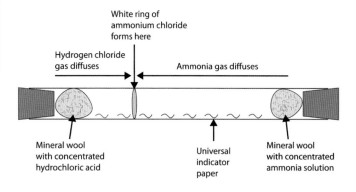

Where the white ring forms:

$$NH_3(g) + HCl(g) \longrightarrow NH_4Cl(s)$$

▲ **Figure 1.12** Hydrochloric acid (left) and ammonia (right) diffuse at different rates

Gases diffuse at different rates. If one piece of cotton wool is soaked in concentrated ammonia solution and another is soaked in concentrated hydrochloric acid and these are put at opposite ends of a dry glass tube, then after a few minutes a white cloud of ammonium chloride appears. Figure 1.12 shows the position at which the two gases meet and react. The white cloud forms in the position shown because the ammonia particles are lighter; they have a smaller relative molecular mass (Chapter 4, p. 52) than the hydrogen chloride particles (released from the hydrochloric acid) and so move faster, such that the gas diffuses more quickly. This experiment is a teacher demonstration only, which must be carried out in a fume cupboard. If considering carrying out this practical, teachers should refer to the *Practical Skills Workbook* for full guidance and safety notes.

Diffusion also takes place in liquids (Figure 1.13) but it is a much slower process than in gases. This is because the particles of a liquid move much more slowly.

(a)   (b)

▲ **Figure 1.13** Diffusion of green food colouring can take days to reach the stage shown in (b)

Diffusion can also take place between a liquid and a gas. Kinetic particle theory can be used to explain this process. It states that collisions are taking place randomly between particles in a liquid or a gas and that there is sufficient space between the particles of one substance for the particles of the other substance to move into.

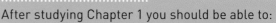

Revision checklist

After studying Chapter 1 you should be able to:
✔ State the three states of matter and describe the structure arrangement of the particles in each.
✔ Explain the properties of the three states of matter using ideas about the separation and movement of particles.
✔ Name the changes of state and describe what happens to the particles in a substance when they occur.
✔ Explain what is happening when a substance changes state.
✔ Describe what happens to a given amount of gas when temperature and/or pressure decreases and increases.
✔ Describe the process of diffusion and explain why gases diffuse.
✔ Explain why the rate of diffusion depends on the molecular mass of the particles.

▶ **Test yourself**

6 When a jar of coffee is opened, people can often smell it from anywhere in the room. Use the kinetic particle theory to explain how this happens.
7 Describe, with the aid of diagrams, the diffusion of a drop of green food coloring added to the bottom of a beaker.
8 Explain why diffusion is faster in gases than in liquids.
9 Explain why a gas with a low relative molecular mass can diffuse faster than a gas with a high relative molecular mass at the same temperature.

Exam-style questions

1 a Sketch diagrams to show the arrangement of particles in:
    i solid oxygen [1]
    ii liquid oxygen [1]
    iii oxygen gas. [1]
  b Describe how the particles move in these three states of matter. [3]
  c Explain, using the kinetic particle theory, what happens to the particles in oxygen as it is cooled down. [3]

2 Explain the meaning of each of the following terms. In your answer include an example to help with your explanation.
  a expansion [2]
  b contraction [2]
  c physical change [2]
  d diffusion [2]
  e random motion [2]

3 a Explain why solids do not diffuse. [2]
  b Give two examples of diffusion of gases and liquids found in your house. [2]

4 Explain the following, using the ideas you have learned about the kinetic particle theory:
  a When you take a block of butter out of the fridge, it is quite hard. However, after 15 minutes it is soft enough to spread. [2]
  b When you come home from school and open the door, you can smell food being cooked. [2]
  c A football is blown up until it is hard on a hot summer's day. In the evening the football feels softer. [2]
  d When a person wearing perfume enters a room, it takes several minutes for the smell to reach the back of the room. [2]

5 Some green food colouring was carefully added to the bottom of a beaker of water using a syringe. The beaker was then covered and left untouched for several days.

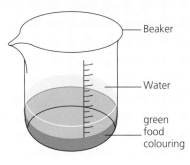

a Describe what you would observe after:
    i a few hours [1]
    ii several days. [1]
  b Explain your answer to Question 5a using your ideas of the kinetic particle theory. [2]
  c State the physical process that takes place in this experiment. [1]

6 The apparatus shown below was set up.

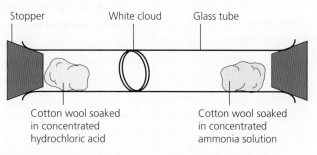

When this apparatus is used, the following things are observed. Explain why each of these is observed.
  a A white cloud is formed. [1]
  b It took a few minutes before the white cloud formed. [1]
  c The white cloud formed further from the cotton wool soaked in ammonia than that soaked in hydrochloric acid. [2]
  d Cooling the concentrated ammonia and hydrochloric acid before carrying out the experiment increased the time taken for the white cloud to form. [1]

Atoms, elements and compounds

FOCUS POINTS
- ★ How are elements, molecules, ions, compounds and mixtures different from each other?
- ★ How do the properties of the particles in an atom lead to an atom's structure?
- ★ What do oxidation and reduction mean?
- ★ What is an isotope?

In Chapter 1 you saw that all matter is made up of particles. In this chapter you will look closely at these particles and see that they are made up of atoms. Atoms are the smallest part of elements. An element is made up of one type of atom and can be either a metal or a non-metal. Metals and non-metals have different properties.

You will look at how atoms of different elements can combine to form substances called compounds, and how this combining occurs in a chemical reaction. By the end of the chapter you should be able to write a simple word or symbol equation to represent these reactions.

You will see that although atoms are the smallest part of an element that shares the chemical properties of that element, they are made from even smaller particles. By learning about the properties and behaviour of these smaller particles (electrons, protons and neutrons), you will be able to see how they affect the chemical properties of elements and compounds.

The universe is made up of a very large number of substances (Figure 2.1), and our own part of the universe is no exception. When we examine this vast array of substances more closely, it is found that they are made up of some basic substances which were given the name **elements** in 1661 by Robert Boyle.

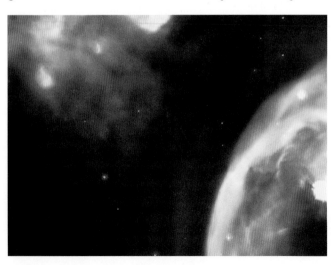

▲ **Figure 2.1** Structures in the universe, such as stars, planets and meteorites, are made of millions of substances. These are made up mainly from just 91 elements, all of which occur naturally on the Earth

In 1803, John Dalton suggested that each element was composed of its own kind of particles, which he called **atoms**. Atoms are much too small to be seen. We now know that about 20×10^6 of them would stretch over a length of only 1 cm.

2.1 Elements

As well as not being able to be broken down into a simpler substance, each element is made up of only one kind of atom. The word atom comes from the Greek word *atomos* meaning 'unsplittable'. For example, aluminium is an element which consists of only aluminium atoms. It is not possible to obtain a simpler substance chemically from the aluminium atoms. You can only combine it with other elements to make more complex substances, such as aluminium oxide, aluminium nitrate or aluminium sulfate.

One hundred and eighteen elements have now been identified. Twenty of these do not occur in nature and have been made artificially by scientists. They include elements such as curium and flerovium. Ninety-eight of the elements occur naturally and range from some very reactive gases, such as fluorine and chlorine, to gold and platinum, which are unreactive elements. A physical property is any characteristic of a substance that we can measure. The elements have different properties that we can measure, and we can then classify them according to those properties.

All elements can be classified according to their various properties. A simple way to do this is to classify them as **metals** or **non-metals** (Figures 2.2 and 2.3). Table 2.1 shows the physical property data for some common metallic and non-metallic elements. You will notice from Table 2.1 that many metals have high densities, high melting points and high boiling points, and that most non-metals have low densities, low melting points and low boiling points. Table 2.2 summarises the different properties of metals and non-metals.

▼ **Table 2.1** Physical data for some metallic and non-metallic elements at room temperature and pressure

Element	Metal or non-metal	Density/ g cm^{-3}	Melting point/°C	Boiling point/°C
Aluminium	Metal	2.70	660	2580
Copper	Metal	8.92	1083	2567
Gold	Metal	19.29	1065	2807
Iron	Metal	7.87	1535	2750
Lead	Metal	11.34	328	1740
Magnesium	Metal	1.74	649	1107
Nickel	Metal	8.90	1453	2732
Silver	Metal	10.50	962	2212
Zinc	Metal	7.14	420	907
Carbon	Non-metal	2.25	Sublimes at 3642	
Hydrogen	Non-metal	0.07[a]	−259	−253
Nitrogen	Non-metal	0.88[b]	−210	−196
Oxygen	Non-metal	1.15[c]	−218	−183
Sulfur	Non-metal	2.07	113	445

Source: Earl B., Wilford L.D.R. Chemistry data book. Nelson Blackie, 1991 a: at −254°C; b: at −197°C; c: at −184°C.

The elements also have chemical properties, which are characteristics or behaviours that may be observed when the substance undergoes a chemical change or reaction. A discussion of the chemical properties of some metals and non-metals is given in Chapters 9 and 10.

a   Gold is very decorative

b   Aluminium has many uses in the aerospace industry

c   These coins contain nickel

▲ **Figure 2.2** Some metals

▼ **Table 2.2** How the properties of metals and non-metals compare

Property	Metal	Non-metal
Physical state at room temperature	Usually solid (occasionally liquid)	Solid, liquid or gas
Malleability	Good	Poor – usually soft or brittle
Ductility	Good	
Appearance (solids)	Shiny (lustrous)	Dull
Melting point	Usually high	Usually low
Boiling point	Usually high	Usually low
Density	Usually high	Usually low
Electrical conductivity and **thermal conductivity**	Good	Very poor

Test yourself

1 Using Tables 2.1 and 2.2, identify one metal in the following list that is different from the others, and explain why it is different.

zinc   copper   oxygen   lead

2 Using Tables 2.1 and 2.2, pick the 'odd one out' in the following group and explain why it is different from the others.

carbon   nitrogen   iron   sulfur

3 Using Tables 2.1 and 2.2, pick the 'odd one out' in the following group of properties of metals and explain why it is different from the others.

- high melting point
- high density
- soft or brittle
- good electrical conductivity

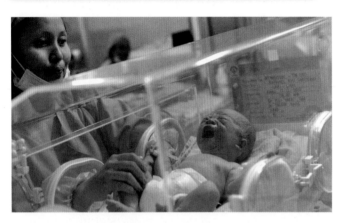

a A premature baby needs oxygen

b Artists often use charcoal (carbon) to produce an initial sketch

c Neon is used in advertising signs

▲ **Figure 2.3** Some non-metals

Atoms – the smallest particles

Everything is made up of billions of extremely small atoms. The smallest atom is hydrogen, and we represent each hydrogen atom as a sphere having a diameter of 0.000 000 07 mm (or 7×10^{-8} mm) (Table 2.3). Atoms of different elements have different diameters as well as different masses.

▼ **Table 2.3** Sizes of atoms

Atom	Diameter of atom/mm	Masses/g
Hydrogen	7×10^{-8}	1.67×10^{-24}
Oxygen	12×10^{-8}	2.66×10^{-23}
Sulfur	20.8×10^{-8}	5.32×10^{-23}

Chemists use symbols to label the elements and their atoms. The symbol comprises one, two or three letters, the first of which is always a capital. The initial letter of the element's name is often used, and where several elements have the same initial letter, another letter from the name is added. For example, **C** is used for carbon, **Ca** for calcium and **Cl** for chlorine. Some symbols seem to have no relationship to the name of the element, for example **Na** for sodium and **Pb** for lead. These symbols come from their Latin names: **na**trium for sodium and **p**lum**b**um for lead. A list of some common elements and their symbols is given in Table 2.4.

▼ **Table 2.4** Some common elements and their symbols. The Latin names of some of the elements are given in brackets

Element	Symbol	Physical state at room temperature and pressure
Aluminium	Al	Solid
Argon	Ar	Gas
Barium	Ba	Solid
Boron	B	Solid
Bromine	Br	Liquid
Calcium	Ca	Solid
Carbon	C	Solid
Chlorine	Cl	Gas
Chromium	Cr	Solid
Copper (Cuprum)	Cu	Solid
Fluorine	F	Gas
Germanium	Ge	Solid
Gold (Aurum)	Au	Solid
Helium	He	Gas
Hydrogen	H	Gas
Iodine	I	Solid
Iron (Ferrum)	Fe	Solid
Lead (Plumbum)	Pb	Solid
Magnesium	Mg	Solid
Mercury (Hydragyrum)	Hg	Liquid
Neon	Ne	Gas
Nitrogen	N	Gas
Oxygen	O	Gas
Phosphorus	P	Solid
Potassium (Kalium)	K	Solid
Silicon	Si	Solid
Silver (Argentum)	Ag	Solid
Sodium (Natrium)	Na	Solid
Sulfur	S	Solid
Tin (Stannum)	Sn	Solid
Zinc	Zn	Solid

Molecules

The atoms of some elements are joined together in small groups. These small groups of atoms are called **molecules**. The atoms of some elements are always joined in pairs, for example, hydrogen, oxygen, nitrogen, fluorine, chlorine, bromine and iodine. They are known as **diatomic** molecules. Using chemical symbols, Cl, the molecule of chlorine shown in Figure 2.4 is written as Cl_2. The atoms of some other elements, such as phosphorus and sulfur, join in larger numbers, four and eight respectively, which we write as P_4 and S_8.

The complete list of the elements with their corresponding symbols is shown in the **Periodic Table** on p. 130.

The gaseous elements helium, neon, argon, krypton, xenon and radon (which are all gases at 0°C at sea level and atmospheric pressure) are composed of separate, individual atoms. When an element exists as separate atoms, then the molecules are said to be **monatomic**. Using chemical symbols, these monatomic molecules are written as He, Ne, Ar, Kr, Xe and Rn, respectively.

Cl —— Cl

a Represented by a letter-and-stick model

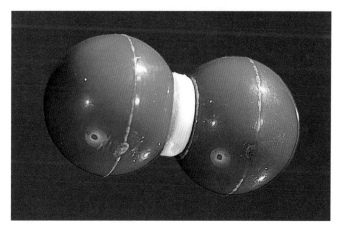

b Represented by a space-filling model

▲ **Figure 2.4** A chlorine molecule

Molecules are not always formed by atoms of the same type joining together as elemental molecules. Most molecules consist of atoms of different elements, for example, water exists as molecules containing oxygen and hydrogen atoms. We will learn more about these in the next section.

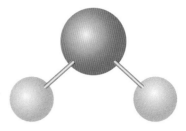

b A model of water showing 2 H atoms and one O atom. Models such as this can be built to show what a compound looks like

▲ **Figure 2.5**

Water molecules contain two atoms of hydrogen and one atom of oxygen, and water has the **molecular formula** H_2O. If there is only one atom of an element in the molecule, no number is required in the formula, only its symbol, as in the case of oxygen in the water molecule H_2O.

> **Key definition**
>
> The **molecular formula** of a compound is defined as the number and type of different atoms in one molecule.

> ▶ **Test yourself**
>
> 4 How many atoms of hydrogen would have to be placed side by side along the edge of your ruler to fill just one of the 1 mm divisions?
> 5 How would you use chemical symbols to write a representation of the molecules of iodine and fluorine?
> 6 Using the Periodic Table on p. 130, write down the symbols for each of these elements and give their physical state at room temperature.
>   a chromium
>   b krypton
>   c osmium

2.2 Compounds

Compounds are pure substances which are formed when two or more elements chemically combine together. A **pure substance** is a material that has a constant composition (is homogeneous) and has consistent properties throughout. Water is a simple compound formed from the elements hydrogen and oxygen (Figure 2.5). This combining of the elements can be represented by a word equation:

hydrogen   +   oxygen   →   water

Hydrogen a pure element	Oxygen a pure element	Hydrogen and oxygen mixed together	Water a pure compound formed from hydrogen burning in oxygen

> ▶ **Test yourself**
>
> 7 What is the formula for the molecule shown in the diagram which contain carbon (black sphere) and hydrogen (white spheres)?

Elements other than hydrogen will also react with oxygen to form compounds called oxides. For example, magnesium reacts violently with oxygen gas to form the white powder magnesium oxide (Figure 2.6). This reaction is accompanied by a release of energy as new chemical bonds are formed.

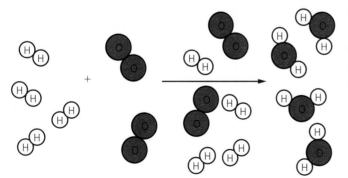

a The element hydrogen reacts with the element oxygen to produce the compound water

Any chemical process that involves reduction and oxidation is known as a **redox reaction**. For example, to extract iron from iron(III) oxide, the oxygen has to be removed. The reduction of iron(III) oxide can be done in a **blast furnace** using carbon monoxide. The iron(III) oxide loses oxygen to the carbon monoxide and is reduced to iron.

Carbon monoxide is the **reducing agent**. A reducing agent is a substance that reduces another substance during a redox reaction. In the reaction carbon monoxide is oxidised to carbon dioxide by the iron(III) oxide. In this process the iron(III) oxide is the **oxidising agent**. An oxidising agent is a substance which oxidises another substance during a redox reaction.

We can write the redox reaction as:

iron(III) oxide + carbon → iron + carbon monoxide dioxide

▲ **Figure 2.6** Magnesium burns brightly in oxygen to produce magnesium oxide

When a new substance is formed during a chemical reaction, a **chemical change** has taken place.

magnesium + oxygen → magnesium oxide

When substances such as hydrogen and magnesium combine with oxygen in this way they are said to have been oxidised, and this process is known as **oxidation**.

Reduction is the opposite of oxidation. In this process oxygen is removed rather than added.

Key definitions

Oxidation is gain of oxygen.

Reduction is loss of oxygen.

Test yourself

8 Zinc is extracted from its ore zinc blende in a furnace by a redox reaction. What does the term 'redox reaction' mean?
9 Identify the oxidising and reducing agents in each of the following reactions:
 a copper(II) oxide + hydrogen → copper + water
 b tin(II) oxide + carbon → tin + carbon dioxide
 c $PbO(s) + H_2(g) \rightarrow Pb(s) + H_2O(l)$

For a further discussion of oxidation and reduction see Chapter 3 (p. 31) and Chapter 5 (p. 68).

Key definitions

Redox reactions involve simultaneous oxidation and reduction.

An **oxidising agent** is a substance that oxidises another substance and is itself reduced.

A **reducing agent** is a substance that reduces another substance and is itself oxidised.

Practical skills

Heating copper

Safety

- Eye protection must be worn.
- Take care when handling hot apparatus.
- Handle the copper with tongs or tweezers, not your fingers.

A student wants to find out what happens when copper is heated in air. In order to do this, they carried out the following experiment and recorded their results.

- First, they found the mass of an empty crucible (a suitably prepared beer-bottle top (metal) is an alternative to a porcelain crucible).
- They added a piece of copper to the crucible and found the mass again.
- They then heated the crucible strongly for approximately two minutes.
- After they allowed it to cool, they then found the mass after heating.

Mass of crucible = 12.90 g
Mass of crucible + copper = 14.18 g
Mass of copper = _____ g
Mass of crucible + contents after heating = 14.30 g

Colour of contents after heating = black

1 Draw a labelled diagram of the experimental set-up used in this experiment.
2 Calculate the change in mass that has taken place during the heating.
3 Explain what has caused the change in mass.
4 What is the black substance left on the copper after heating?
5 Write a word and balanced chemical equation to show the process that has taken place.
6 a How could you modify the experiment to ensure there was no loss of substance taking place during the heating process?
  b What are the other possible sources of error?
7 Predict what would happen, in terms of mass change and colour change, if calcium were heated in air in the same way as the copper.

Formulae

The formula of a compound is made up from the symbols of the elements that make up the compound and numbers that show the ratio of the different atoms from which the compound is made. Carbon dioxide has the formula CO_2, which tells you that it contains one carbon atom for every two oxygen atoms. The 2 in the formula indicates that there are two oxygen atoms present in each molecule of carbon dioxide.

? Worked example

Write the ratio of atoms in sodium sulfate – Na_2SO_4

Substance	Formula	Ratio of atoms
Sodium sulfate	Na_2SO_4	Na : S : O 2 : 1 : 4

▶ Test yourself

10 Write down the ratio of the atoms present in the formula for each of the compounds shown in Table 2.5.

▼ **Table 2.5** Names and formulae of some common compounds

Compound	Formula
Ammonia	NH_3
Calcium hydroxide	$Ca(OH)_2$
Carbon dioxide	CO_2
Copper sulfate	$CuSO_4$
Ethanol (alcohol)	C_2H_5OH
Glucose	$C_6H_{12}O_6$
Hydrochloric acid	HCl
Nitric acid	HNO_3
Sodium carbonate	Na_2CO_3
Sodium hydroxide	$NaOH$
Sulfuric acid	H_2SO_4

The ratio of atoms within a chemical compound is usually constant. Compounds are made up of fixed proportions of elements: they have a fixed composition. Chemists call this the **Law of constant composition**.

For further discussion of formulae, see p. 34.

Balancing chemical equations

Word equations are a useful way of representing chemical reactions, but a better and more useful way of seeing what happens during a chemical reaction is to produce a balanced chemical equation. This type of equation gives the formulae of the substances that are reacting, the reactants, and the new substances formed during the chemical reaction, the products, as well as showing the relative numbers of each of the particles involved.

Balanced equations often include symbols that show the physical state of each of the reactants and products:

» (s) = solid
» (l) = liquid
» (g) = gas
» (aq) = aqueous (water) solution.

We can use the reaction between iron and sulfur as an example. The word equation to represent this reaction is:

$$\text{iron} + \text{sulfur} \xrightarrow{\text{heat}} \text{iron(II) sulfide}$$

When we replace the words with symbols for the reactants and the products, and include their physical state symbols, we get:

$$Fe(s) + S(s) \xrightarrow{\text{heat}} FeS(s)$$

Since there is the same number of each type of atom on both sides of the equation this is called a balanced chemical equation.

? Worked example

Write, for the reaction between magnesium and oxygen producing magnesium oxide:
a   the word equation
b   the balanced chemical equation

a   The word equation is:

$$\text{magnesium} + \text{oxygen} \xrightarrow{\text{heat}} \text{magnesium oxide}$$

b   When we replace the words with symbols for the reactants and the products and include their physical state symbols, it is important to remember that oxygen is a diatomic molecule:

$$Mg(s) + O_2(g) \xrightarrow{\text{heat}} MgO(s)$$

In the equation there are two oxygen atoms on the left-hand side (O_2) but only one on the right (MgO). We cannot

change the formula of magnesium oxide, so to produce the necessary two oxygen atoms on the right-hand side we will need 2MgO – this means 2 × MgO formula units. The equation now becomes:

$$Mg(s) + O_2(g) \xrightarrow{\text{heat}} 2MgO(s)$$

There are now two atoms of magnesium on the right-hand side and only one on the left. To balance the equation, we place a 2 in front of the magnesium, and obtain the following balanced chemical equation:

$$2Mg(s) + O_2(g) \xrightarrow{\text{heat}} 2MgO(s)$$

This balanced chemical equation now shows us that two atoms of magnesium react with one molecule of oxygen gas when heated to produce two units of magnesium oxide.

▶ Test yourself

11 Write the word and balanced chemical equations for the reactions which take place between:
    a   calcium and oxygen
    b   copper and oxygen.

2.3 Mixtures

Many everyday things are not pure substances, they are **mixtures**. A mixture contains more than one substance, which could be elements and/or compounds. Examples of common mixtures are:

» sea water (Figure 2.7)
» air, which is a mixture of elements such as oxygen, nitrogen and neon, and compounds such as carbon dioxide (see Chapter 11, p. 171)
» alloys such as brass, which is a mixture of copper and zinc (for a further discussion of alloys see Chapter 10, p. 159).

▲ **Figure 2.8** The elements sulfur and iron at the top of the photograph, and (below) black iron(II) sulfide on the left and a mixture of the two elements on the right

Substances in a mixture have not undergone a chemical reaction and it is possible to separate them, provided that there is a suitable difference in their physical properties. If the mixture of iron and sulfur is heated, a chemical reaction occurs and a new substance is formed. The product of the reaction is iron(II) sulfide (Figure 2.8 bottom left), and the word equation for this reaction is:

$$\text{iron} + \text{sulfur} \xrightarrow{\text{heat}} \text{iron(II) sulfide}$$

▲ **Figure 2.7** Sea water is a common mixture. It is a water solution of substances such as sodium chloride as well as gases such as oxygen and carbon dioxide

The difference between mixtures and compounds

There are differences between compounds and mixtures, which can be seen by looking at the reaction between iron filings and sulfur. A mixture of iron filings (powdered iron) and sulfur (Figure 2.8 bottom right), looks different from either of the individual elements (Figure 2.8 top). This mixture has the properties of both iron and sulfur; for example, a magnet can be used to separate the iron filings from the sulfur (Figure 2.9).

▲ **Figure 2.9** A magnet will separate the iron from the mixture

During the **reaction** heat energy is released as new chemical bonds are formed. This is called an **exothermic reaction** and accompanies a chemical change (Chapter 6, p. 87). The iron(II) sulfide formed has very different properties to the mixture of iron and sulfur (Table 2.6): for example, iron(II) sulfide would not be attracted towards a magnet. Some chemical reactions take in heat during the reaction, which is called an **endothermic reaction** (Chapter 6, p. 88). You will learn more about the different types of reactions in Chapter 6.

▼ **Table 2.6** Different properties of iron, sulfur, an iron/sulfur mixture and iron(II) sulfide

Substance	Appearance	Effect of a magnet	Effect of dilute hydrochloric acid
Iron	Dark grey powder	Attracted to it	Very little action when cold. When warm, a gas is produced with a lot of bubbling (effervescence)
Sulfur	Yellow powder	None	No effect when hot or cold
Iron/sulfur mixture	Dirty yellow powder	Iron powder attracted to it	Iron powder reacts as above
Iron(II) sulfide	Black solid	No effect	A foul-smelling gas is produced with some effervescence

In iron(II) sulfide, FeS, one atom of iron has combined with one atom of sulfur. In a mixture of iron and sulfur no such ratio exists, as the atoms have not chemically combined. Table 2.7 compares mixtures and compounds. Some common mixtures are discussed in Chapter 10 (p. 159) and Chapter 11 (p. 171).

▼ **Table 2.7** The major differences between mixtures and compounds

Mixture	Compound
It contains two or more substances	It is a single substance
The composition can vary	The composition is always the same
No chemical change occurs when a mixture is formed	When the new substance is formed it involves chemical change
The properties are those of the individual elements/compounds	The properties are very different to those of the component elements
The components may be separated quite easily by physical means	The components can only be separated by one or more chemical reactions

> ## Test yourself
>
> 12 Make a list of some other common mixtures and then use your research to find out and state what they are mixtures of.
>
> 13 Which of the following are not mixtures: milk, tin, sulfur, tap water, brass, gold?

 ## Going further

Other types of mixtures

There are mixtures which are formed by mixing two substances (or phases) which cannot mix. Gels, sols, foams and emulsions are all examples of just such mixtures. Look closely at the substances in Figure 2.10, which shows examples of these different types of mixture.

(a)  (b)

(c) (d)

a This jelly is an example of a gel
b Emulsion paint is an example of a 'sol'

c These foams have been formed by trapping bubbles of gas in liquids or solids
d Emulsions are formed by mixing immiscible liquids

▲ **Figure 2.10**

Composite materials are those that combine the properties of two constituents in order to get the exact properties needed for a particular job. Glass-reinforced fibre is an example of a composite material combining the properties of two different materials. It is made by embedding short fibres of glass in a matrix of plastic. The glass fibres give the plastic extra strength so that it does not break when it is bent or moulded into shape. The finished material has the strength and flexibility of the glass fibres as well as the lightness of plastic (Figure 2.11).

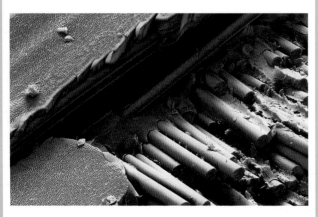

a Glass-reinforced plastic (GRP) consists of glass fibres (rod shapes) embedded in plastic, in this case polyester

b The GRP used to make boats like this is a composite material

▲ **Figure 2.11**

2.4 Inside atoms

Everything you see around you is made out of tiny particles, which we call atoms (Figure 2.12). When John Dalton developed his atomic theory, over 200 years ago, he stated that the atoms of any one element were identical and that each atom was 'indivisible'. Scientists in those days also believed that atoms were solid particles, like marbles.

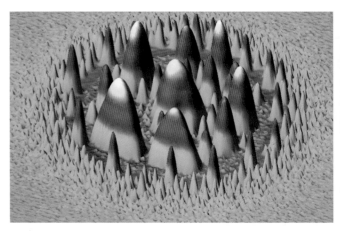

▲ **Figure 2.12** An electron micrograph of atoms taken using a very powerful microscope called an electron microscope

However, in the last hundred years or so it has been proved by great scientists, such as Niels Bohr, Albert Einstein, Henry Moseley, Joseph Thomson, Ernest Rutherford and James Chadwick, that atoms are in fact made up of even smaller 'sub-atomic' particles. Seventy sub-atomic particles have now been discovered, and the most important of these are **electrons**, **protons** and **neutrons**.

These three sub-atomic particles are found in distinct and separate regions of the atom. The protons and neutrons are found in the centre of the atom, which is called the **nucleus**. Neutrons have no charge and protons are positively charged. The nucleus occupies only a very small volume of the atom and is very dense.

The rest of the atom surrounding the nucleus is where electrons are found. Electrons are negatively charged and move around the nucleus very quickly at specific distances from the nucleus in **electron shells** (sometimes called energy levels). The electrons are held in the shells within the atom by an **electrostatic force of attraction** between themselves and the positive charge of protons in the nucleus (Figure 2.13). Each shell can contain only a fixed number of electrons: the first shell can hold up to two electrons, the second shell can hold up to eight electrons, the third shell can hold up to 18, and so on.

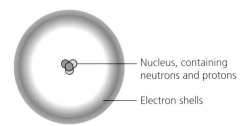

▲ **Figure 2.13** Diagram of an atom

Electrons are tiny and relatively light: approximately 1837 electrons are equal in mass to the mass of one proton or one neutron. A summary of each type of particle, its mass and relative charge is shown in Table 2.8. You will notice that the masses of all these particles are measured in atomic mass units (amu). This is because they are so light that their masses cannot be measured usefully in grams.

▼ **Table 2.8** Characteristics of a proton, a neutron and an electron

Particle	Symbol	Relative mass/ amu	Relative charge
Proton	p	1	+1
Neutron	n	1	0
Electron	e	1/1837	−1

Although atoms contain electrically charged particles, the atoms themselves are electrically neutral (they have no overall electric charge). This is because atoms contain equal numbers of electrons and protons. For example, Figure 2.14 represents the atom of the non-metallic element helium. The atom of helium possesses two protons, two neutrons and two electrons. The electrical charge of the protons in the nucleus is, therefore, balanced by the opposite charge of the two electrons.

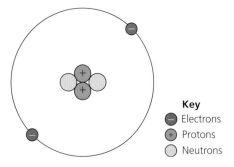

Key
- Electrons
- Protons
- Neutrons

▲ **Figure 2.14** An atom of helium has two protons, two electrons and two neutrons

Proton number and mass number

The number of protons in the nucleus of an atom is called the **proton number** (or atomic number) and is given the symbol Z. The helium atom in Figure 2.14, for example, has a proton number of 2, since it has two protons in its nucleus. Each element has its own proton number and no two elements have the same proton number. For example, the element lithium has a proton number of 3 since it has three protons in its nucleus.

Neutrons and protons have a similar mass, whereas electrons possess very little mass. So the mass of any atom depends on the number of protons and neutrons in its nucleus. The total number of protons and neutrons found in the nucleus of an atom is called the **mass number** (nucleon number) and is given the symbol A.

> **Key definitions**
>
> **Proton number** or atomic number is the number of protons in the nucleus of an atom.
>
> **Mass number** or nucleon number is the total number of protons and neutrons in the nucleus of an atom.

mass number = proton number + number of neutrons
  (A)   (Z)

The helium atom in Figure 2.14 has a mass number of 4, since it has two protons and two neutrons in its nucleus. If we consider the metallic element lithium, it has three protons and four neutrons in its nucleus. It therefore has a mass number of 7.

The proton number and mass number of an element are usually written in the following way:

Mass number (A) $\searrow$ ${}^{4}_{2}He$ $\leftarrow$ Symbol of the element
Proton number (Z) $\nearrow$

The number of neutrons present can be calculated by rearranging the relationship between the proton number, mass number and number of neutrons to give:

number of neutrons = mass number − proton number
                          (A)   (Z)

? Worked example

1   What is the number of neutrons in one atom of ${}^{24}_{12}Mg$?

number of neutrons = mass number − proton number

$12 = 24(A) − 12(Z)$

2   What is the number of neutrons in one atom of ${}^{207}_{82}Pb$?

number of neutrons = mass number − proton number

$125 = 207(A) − 82(Z)$

Table 2.9 shows the number of protons, neutrons and electrons in the atoms of some common elements.

▼ **Table 2.9** Number of protons, neutrons and electrons in some elements

Element	Symbol	Proton number	Number of electrons	Number of protons	Number of neutrons	Mass number
Hydrogen	H	1	1	1	0	1
Helium	He	2	2	2	2	4
Carbon	C	6	6	6	6	12
Nitrogen	N	7	7	7	7	14
Oxygen	O	8	8	8	8	16
Fluorine	F	9	9	9	10	19
Neon	Ne	10	10	10	10	20
Sodium	Na	11	11	11	12	23
Magnesium	Mg	12	12	12	12	24
Sulfur	S	16	16	16	16	32
Potassium	K	19	19	19	20	39
Calcium	Ca	20	20	20	20	40
Iron	Fe	26	26	26	30	56
Zinc	Zn	30	30	30	35	65

Ions

An **ion** is an electrically charged particle. When an atom loses one or more electrons it is no longer electrically neutral and becomes a positively charged ion. This is called a cation.

For example, when potassium is involved in a chemical reaction, each atom loses an electron to form a positive ion, K^+.

            19 protons = 19+
${}_{19}K^+$ <u>18 electrons = 18−</u>
        Overall charge = 1+

When an atom gains one or more electrons it becomes a negatively charged ion. This is called an anion. For example, in some of the chemical reactions involving oxygen, each oxygen atom gains two electrons to form a negative ion, O^{2-}.

          8 protons = 8+
${}_{8}O^{2-}$ <u>10 electrons = 10−</u>
      Overall charge = 2−

The process of gaining or losing electrons is known as **ionisation**.

Table 2.10 shows some common ions. You will notice that:

» some ions contain more than one type of atom, for example NO_3^-
» an ion may possess more than one unit of charge (either negative or positive), for example Al^{3+}, O^{2-} and SO_4^{2-}.

❓ Worked example

1   Show how the sodium ions shown in Table 2.10 are formed.

$$_{11}Na^+ \quad \begin{array}{l} 11 \text{ protons} = 11+ \\ \underline{10 \text{ electrons} = 10-} \\ \text{Overall charge} = \ 1+ \end{array}$$

2   Show how the sulfide ions shown in Table 2.10 are formed.

$$_{16}S^{2-} \quad \begin{array}{l} 16 \text{ protons} \ = 16+ \\ \underline{18 \text{ electrons} \ = 18-} \\ \text{Overall charge} \ = \ 2- \end{array}$$

▼ **Table 2.10** Some common ions

Name	Formula
Lithium ion	Li^+
Sodium ion	Na^+
Potassium ion	K^+
Magnesium ion	Mg^{2+}
Calcium ion	Ca^{2+}
Aluminium ion	Al^{3+}
Zinc ion	Zn^{2+}
Ammonium ion	NH_4^+
Fluoride ion	F^-
Chloride ion	Cl^-
Bromide ion	Br^-
Hydroxide ion	OH^-
Oxide ion	O^{2-}
Sulfide ion	S^{2-}
Carbonate ion	CO_3^{2-}
Nitrate ion	NO_3^-
Sulfate ion	SO_4^{2-}

Test yourself

14 a   State three differences between an electron and a proton.
     b   State two similarities between a neutron and a proton.
     c   Why are atoms electrically neutral?
15 Copy and complete the following table by writing the symbol for the element shown from the given number of particles present. The first one is done for you.

Element	Symbol	Particles present
Nitrogen	$^{14}_{7}N$	7p, 7n, 7e
Aluminium		13p, 14n, 13e
Potassium		19p, 20n, 19e
Argon		18p, 22n, 18e

16 Copy and complete the following table giving the charge on the ions shown.

Element	Proton number	Number of electrons	Charge on ion
Sodium	11	10	
Fluorine	9	10	
Magnesium	12	10	

Isotopes

In some elements, not all of the atoms in a sample of the element are identical. Some atoms of the same element can contain different numbers of neutrons and so have different mass numbers. Atoms of the same element which have different neutron numbers are called **isotopes**. The two isotopes of chlorine are shown in Figure 2.15.

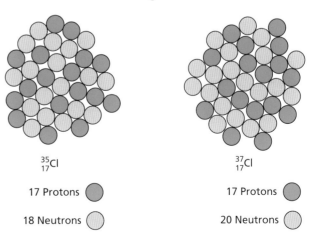

$^{35}_{17}Cl$ $^{37}_{17}Cl$

17 Protons ⬤   17 Protons ⬤

18 Neutrons ◯   20 Neutrons ◯

▲ **Figure 2.15** The nuclei of two isotopes of chlorine

Key definition

Isotopes are different atoms of the same element that have the same number of protons but different numbers of neutrons.

Isotopes of the same element have the same chemical properties because they have the same number of electrons and therefore the same electronic configuration (p. 26).

The only effect of the extra neutron is to alter the mass of the atom and any properties that depend on it, such as density. Other examples of atoms and their isotopes are shown in Table 2.11.

There are two types of isotopes: those which are stable and those which are unstable. The isotopes which are unstable, as a result of the extra neutrons in their nuclei, are **radioactive** and are called **radioisotopes**. The nuclei of these atoms break up spontaneously with the release of not only large amounts of energy, but also certain types of dangerous radiations that can in some cases be useful to society. For example, uranium-235 is used as a source of power in nuclear reactors in nuclear power stations (Chapter 6, p. 87), and cobalt-60 is used in radiotherapy treatment in hospitals (Figure 2.16): these are both radioisotopes.

▼ **Table 2.11** Some atoms and their isotopes

Element	Symbol	Particles present
Hydrogen	$_1^1H$	1e, 1p, 0n
(Deuterium)	$_1^2H$	1e, 1p, 1n
(Tritium)	$_1^3H$	1e, 1p, 2n
Carbon	$_6^{12}C$	6e, 6p, 6n
	$_6^{13}C$	6e, 6p, 7n
	$_6^{14}C$	6e, 6p, 8n
Oxygen	$_8^{16}O$	8e, 8p, 8n
	$_8^{17}O$	8e, 8p, 9n
	$_8^{18}O$	8e, 8p, 10n
Strontium	$_{38}^{86}Sr$	38e, 38p, 48n
	$_{38}^{88}Sr$	38e, 38p, 50n
	$_{38}^{90}Sr$	38e, 38p, 52n
Uranium	$_{92}^{235}U$	92e, 92p, 143n
	$_{92}^{238}U$	92e, 92p, 146n

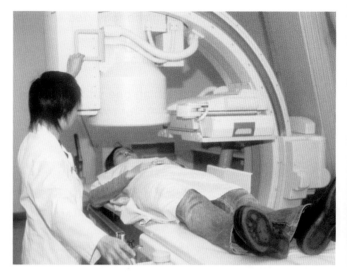

▲ **Figure 2.16** Cobalt-60 is used in radiotherapy treatment. A beam of gamma rays produced by the radioactive isotope is directed into the patient's body to kill tumour tissue.

Relative atomic mass

The average mass of a large number of atoms of an element is called its **relative atomic mass** (A_r). This quantity takes into account the percentage abundance of all the isotopes of an element which exist.

Key definition

Relative atomic mass, A_r, is the average mass of the isotopes of an element compared to 1/12th of the mass of an atom of ^{12}C.

In 1961 the International Union of Pure and Applied Chemistry (IUPAC) recommended that the standard used for the A_r scale was carbon-12. An atom of carbon-12 was taken to have a mass of 12 amu. The A_r of an element is the average mass of the naturally occurring atoms of an element on a scale where ^{12}C has a mass of exactly 12 units.

$$A_r = \frac{\text{average mass of isotopes of the element}}{\frac{1}{12} \times \text{mass of one atom of carbon-12}}$$

Note: $\frac{1}{12}$ of the mass of one carbon-12 atom = 1 amu.

Going further

The mass spectrometer

How do we know isotopes exist? They were first discovered by scientists using apparatus called a mass spectrometer (Figure 2.17). The first mass spectrometer was built by the British scientist Francis Aston in 1919 and enabled scientists to compare the relative masses of atoms accurately for the first time.

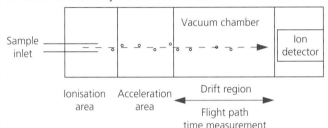

▲ **Figure 2.17** A diagram of a mass spectrometer

A vacuum exists inside the mass spectrometer. A sample of the vapour of the element is injected into the ionisation chamber where it is bombarded by electrons. The collisions which take place between these electrons and the injected atoms cause an electron to be lost from the atom, which becomes a positive ion with a +1 charge. These positive ions are then accelerated towards a negatively charged plate, in the acceleration area.

The spectrometer is set up to ensure that when the ions leave the acceleration area they all have the same kinetic energy, regardless of the mass of the ions. This means that the lighter ions travel faster than the heavier ones, and effectively separate the ions according to their mass. Having left the acceleration area, the time for the ions to reach the detector is recorded. The detector counts the number of each of the ions which fall upon it and so a measure of the percentage abundance of each isotope is obtained. A typical mass spectrum for chlorine is shown in Figure 2.18.

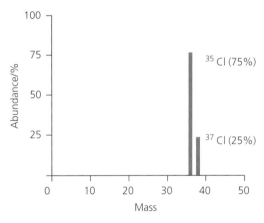

▲ **Figure 2.18** The mass spectrum for chlorine

? Worked example

What is the relative atomic mass of chlorine? Chlorine has two isotopes:

	$^{35}_{17}Cl$	$^{37}_{17}Cl$
% abundance	75	25

Hence the 'average mass' or A_r of a chlorine atom is:

$$\frac{(75 \times 35) + (25 \times 37)}{100} = 35.5$$

$$A_r = \frac{35.5}{1}$$

$$= 35.5 \text{ amu}$$

▶ Test yourself

17 Calculate the number of neutrons in the following atoms:

a $^{27}_{13}Al$ b $^{31}_{15}P$ c $^{91}_{40}Zr$ d $^{190}_{76}Os$

18 Given that the percentage abundance of $^{20}_{10}Ne$ is 90% and that of $^{22}_{10}Ne$ is 10%, calculate the A_r of neon.

The arrangement of electrons in atoms

The nucleus of an atom contains the heavier sub-atomic particles – the protons and the neutrons. The electrons, the lightest of the sub-atomic particles, move around the nucleus at great distances from the nucleus relative to their size. They move very fast in electron shells, very much like the planets orbit the Sun.

It is not possible to give the exact position of an electron in an electron shell. However, we can state that electrons can only occupy certain, definite shells and that they cannot exist between them. Also, as mentioned earlier, each of the electron shells can hold only a certain number of electrons.

» First shell holds up to two electrons.
» Second shell holds up to eight electrons.
» Third shell holds up to 18 electrons.

There are further shells which contain increasing numbers of electrons.

The third shell can be occupied by a maximum of 18 electrons. However, when eight electrons have occupied this shell, a certain stability is given to the atom and the next two electrons go into the fourth shell, and then the remaining ten electrons complete the third shell.

The electrons fill the shells starting from the shell nearest the nucleus, which has the lowest energy. When this is full (with two electrons), the next electron enters the second shell. When this shell is full with eight electrons, then the electrons begin to fill the third and fourth shells as stated above.

For example, a $^{16}_{8}O$ atom has a proton number of 8 and therefore has eight electrons. Two of the eight electrons enter the first shell, leaving six to occupy the second shell, as shown in Figure 2.19. The electronic configuration for oxygen can be written as 2,6.

There are 118 elements, and Table 2.12 shows the way in which the electrons are arranged in the first 20 of these elements. The way in which the electrons are distributed is called the **electronic configuration** (electronic structure). Figure 2.20 shows the electronic configuration of a selection of atoms.

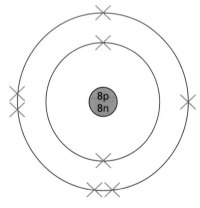

▲ **Figure 2.19** Arrangement of electrons in an oxygen atom

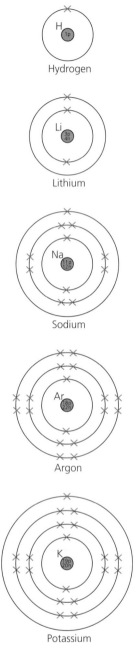

▲ **Figure 2.20** Electronic configurations of hydrogen, lithium, sodium, argon and potassium

▼ **Table 2.12** Electronic configuration of the first 20 elements

Element	Symbol	Proton number	Number of electrons	Electronic configuration
Hydrogen	H	1	1	1
Helium	He	2	2	2
Lithium	Li	3	3	2,1
Beryllium	Be	4	4	2,2
Boron	B	5	5	2,3
Carbon	C	6	6	2,4
Nitrogen	N	7	7	2,5
Oxygen	O	8	8	2,6
Fluorine	F	9	9	2,7
Neon	Ne	10	10	2,8
Sodium	Na	11	11	2,8,1
Magnesium	Mg	12	12	2,8,2
Aluminium	Al	13	13	2,8,3
Silicon	Si	14	14	2,8,4
Phosphorus	P	15	15	2,8,5
Sulfur	S	16	16	2,8,6
Chlorine	Cl	17	17	2,8,7
Argon	Ar	18	18	2,8,8
Potassium	K	19	19	2,8,8,1
Calcium	Ca	20	20	2,8,8,2

From Table 2.12 you can see helium has a full outer shell of two electrons, and neon has a full outer shell of eight electrons. These two elements are very unreactive for this reason. In Chapter 9 we will see that the number of outer shell electrons is related to the position in the Periodic Table. Helium and neon are part of a group of elements known as the noble or inert gases and they are generally very stable and unreactive (p. 140). This is linked to their full outer shells. When elements react to form compounds, they do so to achieve full electron shells, and this idea forms the basis of the electronic theory of chemical bonding, which we will discuss further in the next chapter.

> **Test yourself**
>
> 19 How many electrons may be accommodated in each of the first three shells?
> 20 What is the same about the electronic configurations of:
>   a   lithium, sodium and potassium?
>   b   beryllium, magnesium and calcium?
> 21 An element X has a proton number of 13. What is the electronic configuration of X?

Revision checklist

After studying Chapter 2 you should be able to:

✔ Describe the differences between elements, compounds and mixtures.

✔ Interpret and use symbols for given atoms.

✔ State the formulae of the elements and compounds you have dealt with.

✔ Define the molecular formula of a compound as the number and type of different atoms in one molecule.

✔ Deduce the formula of a simple compound from the relative numbers of atoms present in a model or a diagrammatic representation of the compound.

✔ Construct word equation and symbol equations to show how reactants form products, including state symbols.

✔ Define redox reactions as involving both oxidation and reduction.

✔ Define oxidation as oxygen gain and reduction as oxygen loss.

✔ Identify redox reactions as reactions involving gain and loss of oxygen.

✔ Define an oxidising agent and a reducing agent.

✔ Describe the structure of the atom as a central nucleus containing neutrons and protons surrounded by electrons in shells.

✔ State the relative charges and relative masses of a proton, a neutron and an electron.

✔ Define proton number and atomic number as well as mass number.

✔ Determine the electronic configuration of elements with the proton number 1 to 20.

✔ Describe the formation of positive ions, known as cations, and negative ions, known as anions.

✔ State what isotopes are.

✔ State that isotopes of the same element have the same electronic configuration and so have the same chemical properties.

✔ Calculate the relative atomic mass (A_r) of an element from given data of the relative masses and abundance of their different isotopes.

Exam-style questions

1 a Define the terms:
   i   proton   [3]
   ii neutron   [2]
   iii electron   [3]
   b An atom **X** has a proton number of 19 and
      relative atomic mass of 39.
      i   How many electrons, protons and neutrons
          are there in atom **X**?   [3]
      ii How many electrons will there be in the
          outer electron shell of atom **X**?   [1]
      iii What is the electronic configuration of
          atom **X**?   [1]

2 a $^{69}_{31}Ga$ and $^{71}_{31}Ga$ are isotopes of gallium.
      Use this example to explain what you
      understand by the term isotope.   [3]
   b A sample of gallium is 60% $^{69}_{31}Ga$ atoms
      and 40% $^{71}_{31}Ga$ atoms. Calculate the relative
      atomic mass of this sample of gallium.   [2]

3 Define the following terms using specific
   examples to help with your explanation:
   a element   [2]
   b metal   [3]
   c non-metal   [2]
   d compound   [2]
   e molecule   [2]
   f mixture   [2]

4 State which of the substances listed below are:
   a metallic elements
   b non-metallic elements
   c compounds
   d mixtures.
         silicon, sea water, calcium, argon,
         water, air, carbon monoxide, iron,
         sodium chloride, diamond, brass,
         copper, dilute sulfuric acid, sulfur,
         oil, nitrogen, ammonia   [17]

5 State, at room temperature and pressure (r.t.p.),
   which of the substances listed below is/are:
   a a solid element
   b a liquid element
   c a gaseous mixture
   d a solid mixture
   e a liquid compound
   f a solid compound.
         bromine, carbon dioxide, helium,
         steel, air, oil, marble, copper, water,
         sand, tin, bronze, mercury, salt   [11]

6 a How many atoms of the different elements are
      there in the formulae of these compounds?
      i   nitric acid, HNO_3 [3]
      ii   methane, CH_4 [2]
      iii copper nitrate, $Cu(NO_3)_2$ [3]
      iv   ethanoic acid, CH_3COOH [3]
      v   sugar, $C_{12}H_{22}O_{11}$ [3]
      vi   phenol, C_6H_5OH [3]
      vii ammonium sulfate, $(NH_4)_2SO_4$ [4]
   b Balance the following equations:
      i $Zn(s) + O_2(g) \rightarrow ZnO(s)$ [2]
      ii $Fe(s) + Cl_2(g) \rightarrow FeCl_3(s)$ [3]
      iii $Li(s) + O_2(g) \rightarrow Li_2O(s)$ [2]
      iv $H_2(g) + O_2(g) \rightarrow H_2O(g)$ [2]
      v $Mg(s) + CO_2(g) \rightarrow MgO(s) + C(s)$ [2]

3 Bonding and structure

In the previous chapter you established that both metal and non-metal elements are made up of atoms, which can combine to form compounds. In this chapter you will look more closely at how different elements combine, or bond, to form compounds. You will see that generally when elements combine they can form either ionic bonds with one another and are known as ionic compounds, or they can form covalent bonds with one another and are known as covalent compounds. You will also look at a third type of bonding, found in metals only, called metallic bonding.

You will find out how the type of bonding affects the properties of these compounds. You will also learn more about formulae and how they can be obtained from given data and you will obtain an understanding of redox reactions and the processes associated with them.

3.1 Ionic bonding

Ionic bonds are usually found in compounds that contain metals combined with non-metals. When this type of bond is formed, electrons are transferred from the metal atoms to the non-metal atoms during the chemical reaction. In doing this, the atoms become more stable due to their full outer shells. For example, consider what happens when sodium and chlorine react together and combine to make sodium chloride (Figure 3.1).

sodium + chlorine → sodium chloride

▲ **Figure 3.1** The properties of salt are very different from those of the sodium and chlorine it was made from. To get salt into your diet you would not eat sodium or inhale chlorine!

Sodium has just one electron in its outer shell (Na 2,8,1). Chlorine has seven electrons in its outer shell (Cl 2,8,7). When these two elements react, the outer electron of each sodium atom is transferred to the outer shell of a chlorine atom (Figure 3.2).

In this way both the atoms obtain full outer shells and become 'like' the noble gas nearest to them in the Periodic Table (see Figure 9.3, p. 130). One way of showing this transfer of electrons is by using a dot-and-cross diagram. In these diagrams the electrons from one atom are shown as crosses and those of the other atom as dots. In the reaction the sodium atom has become a sodium ion. This sodium ion has an electronic configuration like the noble gas neon.

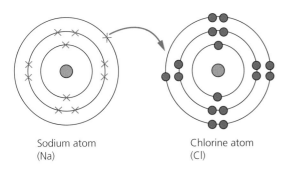

Sodium atom
(Na)

Chlorine atom
(Cl)

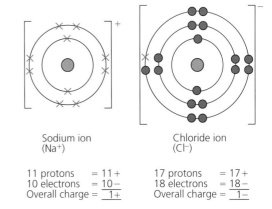

Sodium ion
(Na$^+$)

Chloride ion
(Cl$^-$)

11 protons   = 11+
10 electrons = 10−
Overall charge = 1+

17 protons   = 17+
18 electrons = 18−
Overall charge = 1−

▲ **Figure 3.2** Ionic bonding in sodium chloride

To lose electrons in this way is called **oxidation**. The chlorine atom has become a chloride ion with an electronic configuration like argon. To gain electrons in this way is called **reduction**.

In the chemical process producing sodium chloride both oxidation and reduction have taken place and so this is known as a **redox reaction**.

> **Key definitions**
>
> **Oxidation** involves loss of electrons.
>
> **Reduction** involves gain of electrons.
>
> **Redox reactions** involve simultaneous oxidation and reduction.

A further discussion of oxidation and reduction in terms of electron transfer takes place in Chapter 5 (p. 68).

Only the outer electrons are important in bonding, so we can simplify the diagrams by missing out the inner shells (Figure 3.3).

The charges on the sodium and chloride ions are equal but opposite. They balance each other and the resulting formula for sodium chloride is NaCl. These oppositely charged ions attract each other and are pulled, or bonded, to one another by strong electrostatic forces. This type of bonding is called **ionic bonding**.

> **Key definition**
>
> An **ionic bond** is a strong electrostatic attraction between oppositely charged ions.

All students should be able to describe the formation of ionic bonds between elements of Group I and Group VII and draw dot-and-cross diagrams. Extended learners should be able to do this for any metallic and non-metallic element, including drawing dot-and-cross diagrams.

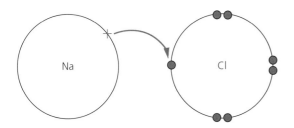

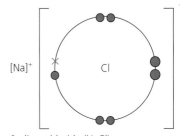

Sodium chloride (NaCl)

▲ **Figure 3.3** Simplified diagram of ionic bonding in sodium chloride

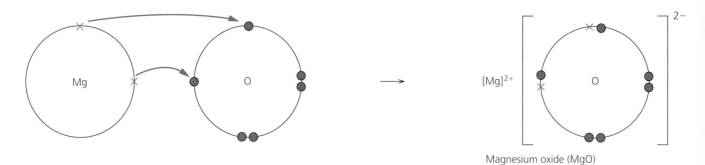

▲ **Figure 3.4** Simplified diagram of ionic bonding in magnesium oxide

Figure 3.4 shows the electron transfers that occur between a magnesium atom and an oxygen atom during the formation of magnesium oxide.

Magnesium obtains a full outer shell by losing two electrons. These are transferred to the oxygen atom. In magnesium oxide, the Mg^{2+} and O^{2-} are oppositely charged and are attracted to one another. The formula for magnesium oxide is MgO.

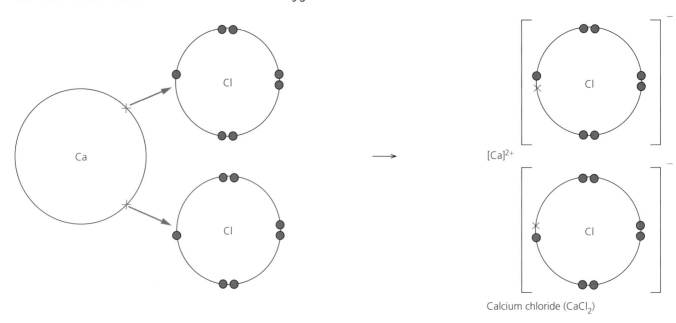

▲ **Figure 3.5** The transfer of electrons that occurs during the formation of calcium chloride

When calcium and chlorine react, the calcium atom gives each of the two chlorine atoms one electron (Figure 3.5). In this case, a compound is formed containing two chloride ions (Cl^-) for each calcium ion (Ca^{2+}). The formula of the compound is $CaCl_2$.

Test yourself

1 Define the term 'ionic bond'.
2 Why do inert (noble) gases *not* bond with themselves and other elements?
3 Draw diagrams to represent the bonding in each of the following ionic compounds:

   a potassium fluoride (KF)
   b lithium chloride (LiCl)
   c magnesium fluoride (MgF_2)
   d calcium oxide (CaO).

Ionic structures

Ionic structures are solids at room temperature and have high melting and boiling points.

The ions are packed together in a regular arrangement called a **lattice**. Within the lattice, oppositely charged ions attract one another strongly.

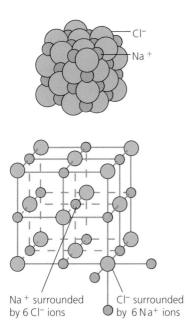

▲ **Figure 3.6** The structure of sodium chloride

Figure 3.6 shows only a tiny part of a small crystal of sodium chloride. Many millions of sodium ions and chloride ions would be arranged in this way in a crystal of sodium chloride to make up the **giant ionic lattice** structure. Each sodium ion in the lattice is surrounded by six chloride ions, and each chloride ion is surrounded by six sodium ions.

 Going further

Not all ionic substances form the same structures. Caesium chloride (CsCl), for example, forms a different structure due to the larger size of the caesium ion compared with that of the sodium ion. This produces the structure shown in Figure 3.7, which is called a body-centred cubic structure. Each caesium ion is surrounded by eight chloride ions and, in turn, each chloride ion is surrounded by eight caesium ions.

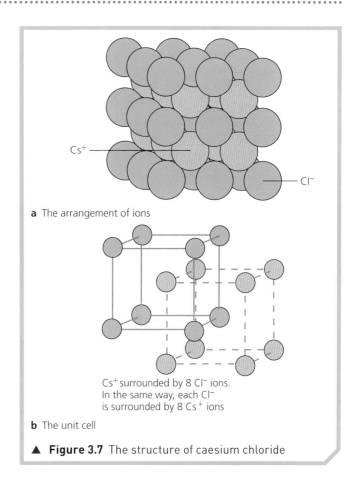

a The arrangement of ions

Cs^+ surrounded by 8 Cl^- ions. In the same way, each Cl^- is surrounded by 8 Cs^+ ions

b The unit cell

▲ **Figure 3.7** The structure of caesium chloride

Properties of ionic compounds

Ionic compounds have the following properties.

>> They are usually solids at room temperature, with high melting and boiling points. This is due to the strong electrostatic forces holding the crystal lattice together. A lot of energy is therefore needed to separate the ions and melt the substance.
>> They are usually hard substances.
>> They mainly dissolve in water.

This is because water molecules are able to bond with both the positive and the negative ions, which breaks up the lattice and keeps the ions apart. Figure 3.8 shows the interaction between water molecules (the solvent) and sodium and chloride ions from sodium chloride (the solute). For a further discussion of the solubility of substances, see Chapter 8, p. 116.

» They usually conduct electricity when in the molten state or in aqueous solution.
The forces of attraction between the ions are weakened and the ions are free to move to the appropriate electrode. This allows an electric current to be passed through the molten compound (see Chapter 5, p. 67).

» They usually cannot conduct electricity when solid, because the ions are not free to move.

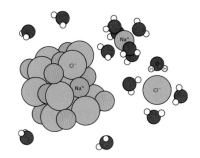

▲ **Figure 3.8** Salt (an ionic compound) dissolving in water

> ### Test yourself
>
> 4   Link the terms in the boxes on the left with the definitions on the right.
>
> | | |
> |---|---|
> | **Oxidation** | An atom or group of atoms which has lost one or more electrons |
> | **Reduction** | Involves loss of electrons |
> | **Redox** | A strong electrostatic force of attraction between oppositely charged ions |
> | **Negative ion** | Involves gain of electrons |
> | **Positive ion** | Involves both oxidation and reduction |
> | **Ionic bond** | An atom or group of atoms which has gained one or more electrons |

Formulae of ionic substances

We saw previously (see p. 31) that ionic compounds contain positive and negative ions, whose charges balance. For example, sodium chloride contains one Na^+ ion for every Cl^- ion, giving rise to the formula NaCl. This method can be used to write down formulae which show the ratio of the number of ions present in any ionic compound.

The formula of magnesium chloride is $MgCl_2$. This formula is reached by each Mg^{2+} ion combining with two Cl^- ions, and once again the charges balance. An oxidation number shows how oxidised or reduced an ion is compared to its atom. Na^+ has an oxidation number of +1 because it is formed by the loss of one electron (Chapter 5, p. 68) from a sodium atom, but Mg^{2+} has an oxidation number of +2 because it is formed when a magnesium atom loses two electrons (Table 3.1). Na^+ can bond (combine) with only one Cl^- ion, whereas Mg^{2+} can bond with two Cl^- ions. A chloride ion has an oxidation state of –1 because it is formed when a chlorine atom gains one extra electron.

Some elements, such as copper and iron, possess two ions with different oxidation numbers. Copper can form the Cu^+ ion and the Cu^{2+} ion, with oxidation numbers 1 and 2 respectively. Therefore it can form two different compounds with chlorine, CuCl and $CuCl_2$. We can also distinguish the difference by using Roman numerals in their names: CuCl is copper(I) chloride and $CuCl_2$ is copper(II) chloride. Similarly, iron forms the Fe^{2+} and Fe^{3+} ions and so can also form two different compounds with, for example, chlorine: $FeCl_2$ (iron(II) chloride) and $FeCl_3$ (iron(III) chloride).

Table 3.1 shows the oxidation numbers of a series of ions you will normally meet in your study of chemistry.

You will notice that Table 3.1 includes groups of atoms which have net charges. For example, the nitrate ion is a single unit composed of one nitrogen atom and three oxygen atoms, and has one single negative charge. The formula, therefore, of magnesium nitrate would be $Mg(NO_3)_2$. You will notice that the NO_3 has been placed in brackets with a $_2$ outside the bracket. This indicates that there are two nitrate ions present for every magnesium ion. The ratio of the atoms present is therefore:

$$Mg(NO_3)_2$$

$$1Mg : 2N : 6O$$

▼ **Table 3.1** Oxidation numbers (valencies) of some elements (ions) and groups of atoms

	Oxidation number (valency)					
	1		**2**		**3**	
Metals	Lithium	(Li^+)	Magnesium	(Mg^{2+})	Aluminium	(Al^{3+})
	Sodium	(Na^+)	Calcium	(Ca^{2+})	Iron(III)	(Fe^{3+})
	Potassium	(K^+)	Copper(II)	(Cu^{2+})		
	Silver	(Ag^+)	Zinc	(Zn^{2+})		
	Copper(I)	(Cu^+)	Iron(II)	(Fe^{2+})		
			Lead	(Pb^{2+})		
			Barium	(Ba^{2+})		
Non-metals	Fluoride	(F^-)	Oxide	(O^{2-})		
	Chloride	(Cl^-)	Sulfide	(S^{2-})		
	Bromide	(Br^-)				
	Hydrogen	(H^+)				
Groups of atoms	Hydroxide	(OH^-)	Carbonate	(CO_3^{2-})	Phosphate	(PO_4^{3-})
	Nitrate	(NO_3^-)	Sulfate	(SO_4^{2-})		
	Ammonium	(NH_4^+)	Dichromate(VI)	$(Cr_2O_7^{2-})$		
	Hydrogencarbonate	(HCO_3^-)				
	Manganate(VII)	(MnO_4^-)				

Oxidation number

Each atom in an element or compound is assigned an oxidation number to show how much it is reduced or oxidised. The following points should be remembered when using oxidation numbers.

» Roman numerals (I, II, III, IV, V, VI, VII, VIII) are used in writing the oxidation number of an element.
» This number is placed after the element that it refers to. For example, the name for $FeCl_3$ is iron(III) chloride and not iron(3) chloride.
» The oxidation number of the free element is always 0, for example in metals such as zinc and copper.
» In simple monatomic ions, the oxidation number is the same as the charge on the ion. So iodine has an oxidation number of 0 in I_2 but an oxidation number of –1 in I^-.
» Compounds have no charge overall. Hence the oxidation numbers of all the individual elements in a compound must add up to 0. The oxidation numbers of elements in compounds can vary, as seen in Table 3.1. It is possible to recognise which of the different oxidation numbers a metal element is in by the colour of its compounds (Figure 3.9).

▲ **Figure 3.9** Iron(II) sulfate is pale green, while iron(III) sulfate is yellow

» An increase in the oxidation number, for example from +2 to +3 as in the case of Fe^{2+} to Fe^{3+}, is oxidation.

>> However, a reduction in the oxidation number, for example from +6 to +3 as in the case of Cr^{6+} (in CrO_4^{2-}) to Cr^{3+}, is reduction.

> **Key definitions**
>
> **Oxidation** involves an increase in oxidation number.
>
> **Reduction** involves a decrease in oxidation number.

During a redox reaction, the substance that brings about **oxidation** is called an **oxidising agent** and is itself reduced during the process. A substance that brings about **reduction** is a **reducing agent** and is itself oxidised during the process.

For example, if a dilute solution of acidified potassium manganate(VII) (pale purple) is added to a solution of iron(II) sulfate, a colour change takes place as the reaction occurs (Figure 3.10). The iron(II) sulfate (pale green) changes to pale yellow, showing the presence of iron(III) ions.

▲ **Figure 3.10** Manganate(VII) ions (oxidising agent) and iron(II) ions (reducing agent) are involved in a redox reaction when mixed

In this reaction the iron(II) ions have been oxidised to iron(III) ions (increase in oxidation number) and the manganate(VII) ions have been reduced to manganese(II) ions (decrease in oxidation number), which are very pale pink. Hence the manganate(VII) ions are the oxidising agent and the iron(II) ions are the reducing agent.

Potassium iodide is a common reducing agent. When it is oxidised, a colour change is also produced. The iodide ion (I^-) is oxidised to iodine(I_2). The colour of the resulting solution will change from colourless to yellow-brown. If you then

add starch indicator, a test to show the presence of iodine, it will turn blue-black. Hence if potassium iodide solution is added to a solution of iron(III) solution followed by the starch indicator, then a blue-black colour will be seen as shown in Figure 3.11. This shows that the iron(III) ions must have been reduced to iron(II) ions as the iodide ions are oxidised to iodine.

> **Key definitions**
>
> An **oxidising agent** is a substance that oxidises another substance and is itself reduced.
>
> A **reducing agent** is a substance that reduces another substance and is itself oxidised.

▲ **Figure 3.11** The blue-black colour shows the presence of iron(III) ions

It is possible to recognise redox processes by looking at the oxidation numbers on the two sides of the chemical equation for a reaction. For example, in the reaction between magnesium and dilute sulfuric acid, the magnesium dissolves and hydrogen gas is produced. Both magnesium metal and hydrogen gas are free elements and so have an oxidation number of 0. In sulfuric acid, hydrogen has an oxidation number of +1 since the overall charge on the sulfate ion is −2. Similarly, the oxidation number of magnesium in magnesium sulfate is +2.

magnesium + sulfuric acid → magnesium sulfate + hydrogen

$$Mg(s) \quad + H_2SO_4(aq) \rightarrow \quad MgSO_4(aq) \quad + \quad H_2(g)$$
$$\;\;0 \qquad\quad +1 \qquad\qquad\quad +2 \qquad\qquad\qquad 0$$

The sulfate ions are unchanged by the reaction and so can be ignored.

As you can see, the oxidation number of magnesium has increased from 0 to +2. Therefore the magnesium has been oxidised by the sulfuric acid and so sulfuric acid is the **oxidising agent**. The oxidation number of hydrogen has decreased from +1 in the sulfuric acid to 0 in the free element. Therefore the hydrogen has been reduced by the magnesium and so magnesium is the **reducing agent**.

> **Test yourself**
>
> 5 Identify the oxidising and reducing agents in the following reactions:
> a $Zn(s) + H_2SO_4(aq) \rightarrow ZnSO_4(aq) + H_2(g)$
> b $Cu^{2+}(aq) + Zn(s) \rightarrow Cu(s) + Zn^{2+}(aq)$
> c $Mg(s) + Cu(NO_3)_2(aq) \rightarrow Mg(NO_3)_2(aq) + Cu(s)$
> d $Fe(s) + H_2SO_4(aq) \rightarrow FeSO_4(aq) + H_2(g)$
> 6 Using the information in Table 3.1, write the formulae for:
> a copper(I) oxide
> b zinc phosphate
> c iron(III) chloride
> d lead(II) bromide
> e potassium manganate(VII)
> f sodium dichromate(VI)
> 7 Using the formulae in your answer to question **6**, write down the ratio of atoms present for each of the compounds.

> ➡ **Going further**
>
> **The 'cross-over' method**
>
> A less scientific but simpler method to work out the formulae of compounds is called the 'cross-over' method. In this method it is only necessary to 'swap' the numerical value of the oxidation number of the elements or groups of atoms (or radicals) concerned. This is easily done by 'crossing over' the numerical value of the oxidation number, by placing the numerical value of the oxidation number of the first element after the symbol of the second and placing the numerical value of the oxidation number of the second element or radical after the symbol of the first.
>
> For example, in aluminium sulfide the oxidation numbers of the elements are:
>
> $Al = 3$ and $S = 2$
>
> Hence the formula of the compound is Al_2S_3.
>
> Similarly, in sodium sulfate the oxidation numbers are:
>
> $Na = 1$ and $SO_4 = 2$
>
> Hence the formula of the compound is Na_2SO_4.

3.2 Covalent bonding

Another way in which atoms can form stable compounds is by sharing the electrons in their outer shells. This occurs between non-metal atoms, and the bond formed is called a **covalent bond**. During the bond formation the atoms involved gain the stability of the noble (inert) gas electronic configuration. The simplest example of this type of bonding can be seen by considering the hydrogen molecule, H_2.

> **Key definition**
>
> A **covalent bond** is formed when a pair of electrons is shared between two atoms leading to noble gas electronic configurations.

Each hydrogen atom in the molecule has one electron. In order to obtain a full outer shell and gain an electronic configuration that is the same as the noble gas helium, each of the hydrogen atoms must have two electrons. To do this, the outer shells of the two hydrogen atoms overlap (Figure 3.12a). A molecule of hydrogen is formed, with two hydrogen atoms sharing a pair of electrons (Figure 3.12a and b). This shared pair of electrons is known as a single covalent bond and is represented by a single line as in hydrogen:

H—H

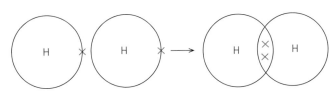

2 hydrogen atoms   Hydrogen molecule (H_2)

a The electron sharing to form the single covalent bond in H_2 molecules

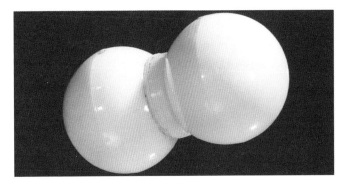

b Model of a hydrogen molecule

▲ **Figure 3.12**

A similar example exists in the diatomic halogen molecule chlorine, Cl_2 (Figure 3.13).

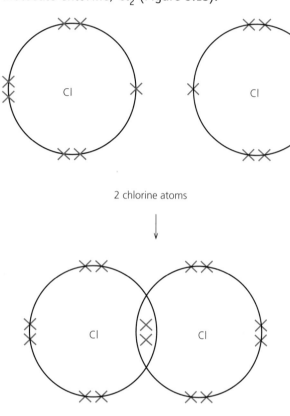

2 chlorine atoms

Chlorine molecule (Cl_2)

▲ **Figure 3.13** The electron sharing to form the single covalent bond in Cl_2 molecules (showing outer electron shells only)

Another simple molecule is hydrogen chloride. This time two different elements share electrons to gain the electronic configuration of their nearest noble gas. In this case hydrogen, with only one electron in its outer shell, needs to share one more electron to gain the electronic configuration of helium. Chlorine has 7 electrons in its outer shell so it also needs to share one more electron, this time to gain the electronic configuration of the noble gas argon (Figure 3.14).

▲ **Figure 3.14** Dot-and-cross diagram to show the formation of a hydrogen chloride molecule

Other covalent compounds

Methane (natural gas) is a gas whose molecules contain atoms of carbon and hydrogen. The electronic configurations are:

C 2,4   H 1

The carbon atom needs four more electrons to attain the electronic configuration of the noble gas neon. Each hydrogen atom needs only one electron to form the electronic configuration of helium. Figure 3.15 shows how the atoms gain these electronic configurations by the sharing of electrons. Note that only the outer electron shells are shown. Figure 3.16 shows a model of a methane molecule.

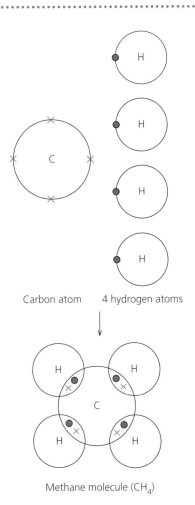

Carbon atom   4 hydrogen atoms

Methane molecule (CH₄)

▲ **Figure 3.15** Dot-and-cross diagram to show the formation of a methane molecule

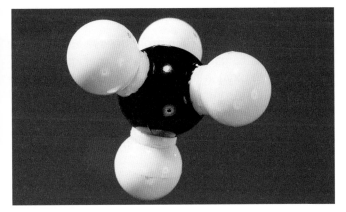

▲ **Figure 3.16** Model of a methane molecule

Ammonia is a gas containing the elements nitrogen and hydrogen. It is used in large amounts to make fertilisers. The electronic configurations of the two elements are:

N 2,5   H 1

The nitrogen atom needs three more electrons to obtain the noble gas structure of neon. Each hydrogen requires only one electron to form the noble gas structure of helium. The nitrogen and hydrogen atoms share electrons, forming three single covalent bonds (Figure 3.17). Figure 3.18 shows a model of an ammonia molecule.

Water is a liquid containing the elements hydrogen and oxygen. The electronic configurations of the two elements are:

O 2,6   H 1

The oxygen atom needs two electrons to gain the electronic configuration of neon. Each hydrogen requires one more electron to gain the electronic configuration of helium. Again, the oxygen and hydrogen atoms share electrons, forming a water molecule with two single covalent bonds as shown in Figure 3.19. Figure 3.20 shows a model of a water molecule.

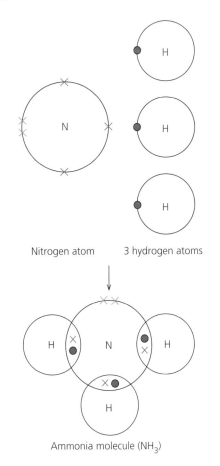

Nitrogen atom   3 hydrogen atoms

Ammonia molecule (NH₃)

▲ **Figure 3.17** Dot-and-cross diagram to show the formation of an ammonia molecule

▲ **Figure 3.18** Model of the ammonia molecule

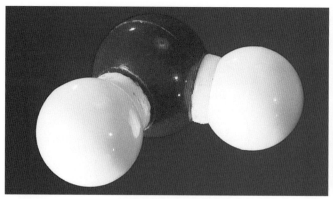

▲ **Figure 3.20** Model of a water molecule – this is a V-shaped molecule

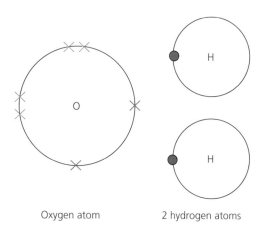

Oxygen atom   2 hydrogen atoms

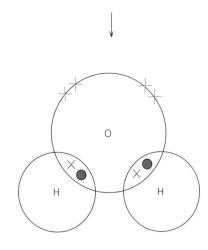

Water molecule (H_2O)

▲ **Figure 3.19** Dot-and-cross diagram to show the formation of a water molecule

Going further

The shapes of molecules

The methane molecule has a tetrahedral shape (Figure 3.16). This shape is caused by the repulsion of the four C–H bonding pairs of electrons. The electron pairs arrange themselves as far apart as possible.

The pyramidal shape of the ammonia molecule (Figure 3.18) is caused by the repulsion between the bonding pairs of electrons in each of the three N–H bonds, as well as the non-bonding pair of electrons.

Carbon dioxide is a gas containing the elements carbon and oxygen. The electronic configurations of the two elements are:

C 2,4   O 2,6

In this case each carbon atom needs to share four electrons to gain the electronic configuration of neon. Each oxygen needs to share two electrons to gain the electronic configuration of neon. This is achieved by forming two double covalent bonds in which two pairs of electrons are shared in each case, as shown in Figure 3.21. Carbon dioxide is a linear molecule (Figure 3.22).

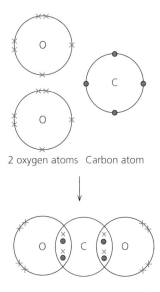

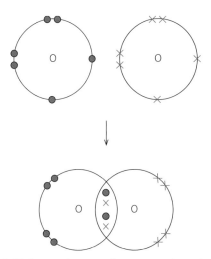

2 oxygen atoms   Carbon atom

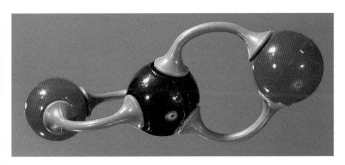

Carbon dioxide molecule (CO₂)

▲ **Figure 3.21** Dot-and-cross diagram to show the formation of a carbon dioxide molecule

o═c═o

a Carbon dioxide molecule. Note the double covalent bond is represented by a double line

b Model of the linear carbon dioxide molecule

▲ **Figure 3.22**

Another molecule which contains a double covalent bond is that of oxygen gas, O_2 (Figure 3.23). Each of the oxygen atoms has six electrons in their outer electron shell.

O 2,6

Each needs to share another two electrons to gain the electronic configuration of the noble gas neon.

▲ **Figure 3.23** Dot-and-cross diagram to show the formation of an oxygen molecule

As with the carbon dioxide molecule the double bond can be shown by a double line between the two oxygen atoms, representing two shared pairs of electrons.

O=O

Nitrogen gas, N_2, is molecule which contains a triple covalent bond (Figure 3.24). A triple covalent bond is formed when three pairs of electrons are shared and it is represented by '≡'.

Each nitrogen atom has the electronic configuration shown below:

N 2,5

Both nitrogen atoms in the molecule need to share a further three electrons to gain the electronic configuration of the noble gas neon.

The triple bond can be shown in the molecule as N≡N.

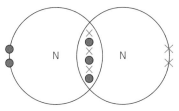

▲ **Figure 3.24** Dot-and-cross diagram to show the bonding in a nitrogen molecule

Methanol, CH_3OH, is a molecule which contains three different types of atom (Figure 3.25). When this happens and you need to draw a dot-and-cross diagram, simply make sure that atoms which bond with one another do not both have dots or crosses. In this molecule the atoms have these configurations:

C 2,4   H 1   O 2,6

The carbon atom needs to share four further electrons which it can do with the three hydrogens: each needs to share one more electron, and the oxygen atom. The oxygen atom, which shares one electron with the carbon atom, shares another electron with the remaining hydrogen atom. By doing this the hydrogen atoms gain the electronic configuration of helium and the carbon and oxygen atoms gain the electronic configuration of neon.

The dot-and-cross diagram for methanol is:

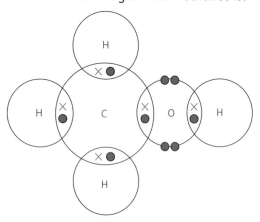

▲ **Figure 3.25** Dot-and-cross diagram to show the formation of a methanol molecule

> ### Test yourself
>
> 8   What do you understand by the term 'covalent bond'?
> 9   Draw diagrams to represent the bonding in each of the following covalent compounds:
>   a   tetrachloromethane   b   hydrogen sulfide
>   (CCl_4)   (H_2S)

Covalent structures

Compounds containing covalent bonds have molecules whose structures can be classified as either **simple molecular** or **giant covalent**.

Simple molecular structures are simple, formed from only a few atoms. They have strong covalent bonds between the atoms within a molecule but have weak bonds between the molecules (**intermolecular forces**).

> **Going further**
>
> Some of the strongest of these weak intermolecular forces occur between water molecules.
>
> One type of weak bond between molecules is known as the van der Waals bond (or force), and these forces increase steadily with the increasing size of the molecule.

Examples of simple molecules are iodine (Figure 3.26), methane, water and ethanol.

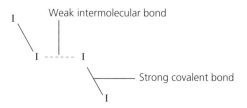

▲ **Figure 3.26** Strong covalent and weak intermolecular forces in iodine

Giant covalent structures contain many hundreds of thousands of atoms joined by strong covalent bonds. Examples of substances with this type of structure are diamond, graphite and silicon(IV) oxide (Figure 3.27).

Properties of covalent compounds

Covalent compounds have the following properties.

» As simple molecular compounds, they are usually gases, liquids or solids with low melting and boiling points. The melting points are low because of the weak intermolecular forces of attraction which exist between simple molecules. These are weaker compared to the strong covalent bonds. Giant covalent substances have higher melting points, because the whole structure is held together by strong covalent bonds.
It should be noted that in ionic compounds, the interionic forces are much stronger than the intermolecular forces in simple covalent substances and so the melting and boiling points are generally higher.

» Generally, they do not conduct electricity when molten or dissolved in water. This is because they do not contain ions. However, some molecules react with water to form ions. For example, hydrogen chloride gas produces aqueous hydrogen ions and chloride ions when it dissolves in water:

$$HCl(g) \xrightarrow{\text{water}} H^+(aq) + Cl^-(aq)$$

The presence of the ions allows the solution to conduct electricity.

» Generally, they do not dissolve in water. However, water is an excellent solvent and can interact with and dissolve some covalent molecules better than others. Covalent substances are generally soluble in organic solvents.

 Going further

Giant covalent structures

When an element can exist in more than one physical form in the same physical state, it is said to exhibit **allotropy** (or polymorphism). Each of the different physical forms is called an allotrope. Allotropy is actually quite a common feature of the elements in the Periodic Table. An example of an element which shows allotropy is carbon.

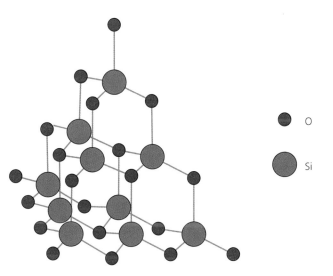

O
Si

a The silicon(IV) oxide structure in quartz

b Quartz is a hard solid at room temperature. It has a melting point of 1610°C and a boiling point of 2230°C

▲ **Figure 3.27**

Practical skills

Testing the electrical conductivity of ionic and covalent substances

Safety

- Eye protection must be worn.
- Take care when handling dilute aqueous solutions: most are low hazard (and should all be ≤1M); 0.2–1.0 mol/dm³ copper(II) sulfate solution is an irritant and corrosive (Hazard).

The apparatus was set up as shown in the diagram to investigate the electrical conductivity of a number of compounds in aqueous solution.

1 Describe the experimental method you could use to carry out a fair test on a variety of ionic

and covalent substances dissolved in water and using the apparatus shown in the diagram.

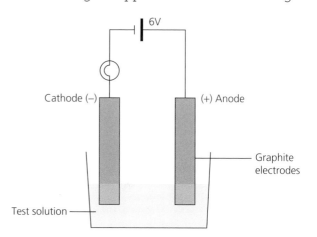

Solution	Does the bulb light?	Does the substance in solution conduct electricity?	Does the substance in solution contain ionic or covalent bonding?
Glucose ($C_6H_{12}O_6$)	No		
Potassium chloride	Yes		
Calcium chloride	Yes		
Sodium iodide	Yes		
Ethanol (C_2H_5OH)	No		
Copper(II) sulfate	Yes		
Pure water	No		

2 Copy and complete the results table above.
3 Using your answers to columns three and four in the results table, write a conclusion for this experiment. In your conclusion, state and explain any trends shown in the results.
4 Would the results be different if solid potassium chloride was used instead of potassium chloride solution? Explain your answer.

5 Predict and explain the results you would obtain for magnesium nitrate solution.
6 Predict and explain the results you would obtain with just distilled water – that is, no dissolved substance present.

Different forms of carbon

Carbon is a non-metallic element which exists in more than one solid structural form. These are graphite and diamond. Each of the forms has a different structure (Figures 3.28 and 3.29) and

so they exhibit different physical properties (Table 3.2). These different physical properties lead to graphite and diamond being used in different ways (Table 3.3 and Figure 3.30).

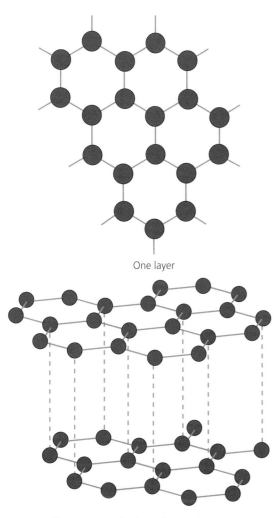

One layer

Showing how the layers fit together

a A portion of the graphite structure

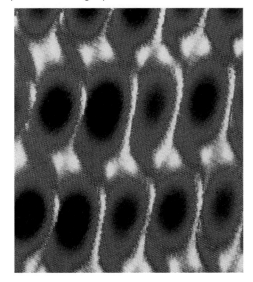

b A piece of graphite as imaged through a scanning tunnelling microscope

▲ **Figure 3.28**

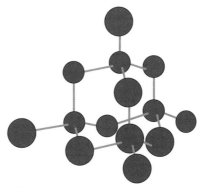

A small part of the structure

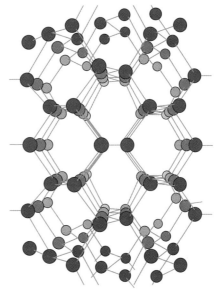

A view of a much larger part of the structure

a The structure of diamond

b A 100-carat, flawless diamond

▲ **Figure 3.29**

▼ **Table 3.2** Physical properties of graphite and diamond

Property	Graphite	Diamond
Appearance	A dark grey, shiny solid	A colourless transparent crystal which sparkles in light
Electrical conductivity	Conducts electricity	Does not conduct electricity
Hardness	A soft material with a slippery feel	A very hard substance
Density (g/cm³)	2.25	3.51

▼ **Table 3.3** Uses of graphite and diamond

Graphite	Diamond
Pencils	Jewellery
Electrodes	Glass cutters
Lubricant	Diamond-studded saws
	Drill bits
	Polishers

Graphite

Figure 3.28a shows the structure of graphite. This is a layer structure. Within each layer each carbon atom is bonded to three others by strong covalent bonds. Each layer is therefore like a giant molecule. Between these layers there are weak forces of attraction and so the layers will pass over each other easily.

With only three covalent bonds formed between carbon atoms within the layers, an unbonded electron is present on each carbon atom. These 'spare' (or **delocalised**) electrons move freely between the layers and it is because of these spare electrons that graphite conducts electricity.

 Going further

Graphitic compounds

In recent years a set of interesting compounds known as graphitic compounds have been developed. In these compounds, different atoms have been fitted in between the layers of carbon atoms to produce a substance with a greater electrical conductivity than pure graphite. Graphite is also used as a component in certain sports equipment, such as tennis and squash rackets.

 Going further

Graphene

Discovered in 2004, graphene is a so-called super material made up of single layers of graphite as shown in the upper diagram of Figure 3.28. It is able to conduct electricity one million times better than copper metal and has enormous potential in electronics.

Diamond

Figure 3.29 shows the diamond structure. Each of the carbon atoms in the giant structure is covalently bonded to four others. They form a tetrahedral arrangement similar to that found in silicon(IV) oxide (p. 43). This bonding scheme produces a very rigid three-dimensional structure and accounts for the extreme hardness of the substances silicon(IV) oxide and diamond. All the outer shell electrons of the carbon atoms are used to form covalent bonds, so there are no electrons available to enable diamond to conduct electricity.

It is possible to manufacture the different allotropes of carbon. Diamond is made by heating graphite to about 300°C at very high pressures. Diamond made by this method is known as industrial diamond. Graphite can be made by heating a mixture of coke and sand at a very high temperature in an electric arc furnace for about 24 hours.

The various uses of graphite and diamond result from their differing properties (Figure 3.30).

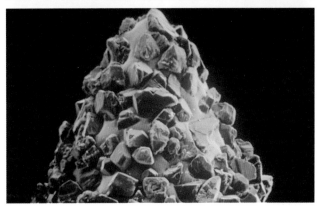

▲ **Figure 3.30** Uses of graphite (as a pencil 'lead' and as a lock lubricant) and diamond (as a toothed saw to cut marble and on a dentist's drill)

➡ Going further

Buckminsterfullerene – an unusual form of carbon

In 1985, a new allotrope of carbon was obtained by Richard Smalley and Robert Curl of Rice University, Texas. It was formed by the action of a laser beam on a sample of graphite.

The structure of buckminsterfullerene can be seen in Figure 3.31.

This spherical structure is composed of 60 carbon atoms covalently bonded together. Further spherical forms of carbon, 'bucky balls', containing 70, 72 and 84 carbon atoms have been identified and the discovery has led to a whole new branch of inorganic carbon chemistry. It is thought that this type of molecule exists in chimney soot. Chemists have suggested that, due to the large surface area of the bucky balls, they may have uses as catalysts (Chapter 7, p. 102). Also they may have uses as superconductors.

Buckminsterfullerene is named after an American architect, Buckminster Fuller, who built complex geometrical structures (Figure 3.32).

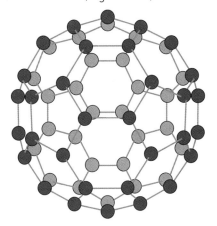

▲ **Figure 3.31** Buckminsterfullerene – a 'bucky ball' (C_{60})

▲ **Figure 3.32** C$_{60}$ has a structure similar to a football and to the structure of the domes shown

 Test yourself

10 Explain the difference between ionic and covalent bonding.

11 Compare the structures of silicon(IV) oxide and diamond.

12 Draw up a table to summarise the properties of the different types of substances you have met in this chapter. Your table should include examples from ionic substances and covalent substances (simple and giant).

Going further

Glasses and ceramics

Glasses

Glass can be made by heating silicon(IV) oxide with other substances until a thick viscous liquid is formed. As this liquid cools, the atoms present cannot move freely enough to return to their arrangement within the pure silicon(IV) oxide structure. Instead they are forced to form a disordered arrangement as shown in Figure 3.33. Glass is called a supercooled liquid because it has cooled below its freezing point without solidification into a crystalline structure. There are other types of disordered substance formed this way, called glasses, which are irregular giant molecular structures held together by strong covalent bonds.

Ceramics

The word *ceramic* comes from the Greek word *keramos* meaning pottery or 'burnt stuff'. Clay dug from the ground contains a mixture of several materials including silicon(IV) oxide. During firing in a furnace, the clay is heated to a temperature of 1000°C. The material produced at the end of the firing, the ceramic, which is a type of glass, consists of many tiny mineral crystals bonded together with glass.

Other modern ceramics include zirconium oxide and silicon nitride. There are now many uses of these

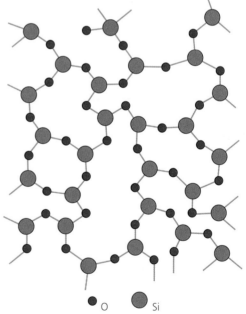

O ● Si ●

▲ **Figure 3.33** Two-dimensional structure of silicon(IV) oxide in glass

ceramic materials, such as ceramic discs, which prevent short-circuits in the national grid pylons; ceramic tiles, which protect the spacecraft; and ceramic bearings that do not need lubrication.

3.3 Metallic bonding

Another way in which atoms obtain a more stable electronic configuration is found in metals. The electrons in the outer shell of the atom of a metal move freely throughout the structure. They are delocalised, forming a mobile 'sea' of electrons (Figure 3.34). When the metal atoms lose these electrons, they form a giant lattice of positive ions. Therefore, metals consist of positive ions embedded in a sea of moving electrons. The negatively charged electrons attract all the positive metal ions and bond them together with strong electrostatic forces of attraction as a single unit. This is the **metallic bond**.

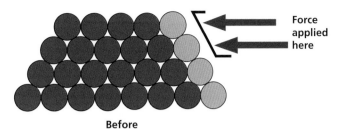

▲ **Figure 3.35** The positions of the positive ions in a metal before and after a force has been applied

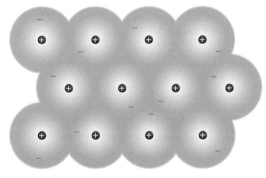

▲ **Figure 3.34** Metals consist of positive ions surrounded by a 'sea' of delocalised electrons

Properties of metals

Metals have the following properties.

» They conduct electricity due to the mobile electrons within the metal structure. When a metal is connected in a circuit, the electrons move towards the positive terminal while at the same time electrons are fed into the other end of the metal from the negative terminal.

» They are malleable and ductile. Unlike the fixed bonds in diamond, metallic bonds are not rigid, but they are still strong. If a force is applied to a metal, rows of ions can slide over one another. They reposition themselves and the strong bonds re-form as shown in Figure 3.35. **Malleable** means that metals can be bent or hammered into different shapes. **Ductile** means that the metals can be pulled out into thin wires.

» They usually have high melting and boiling points due to the strong attraction between the positive metal ions and the mobile 'sea' of electrons.

» They have high densities because the atoms are very closely packed in a regular manner, as can be seen in Figure 3.36. Different metals have different types of packing of atoms and in doing so they produce the arrangements of ions shown in Figure 3.37.

▲ **Figure 3.36** Arrangement of ions in the crystal lattice of a metal

> ## Test yourself
> 13 Explain the terms:
>   a   malleable
>   b   ductile
> 14 Explain why metals are able to conduct electricity.

Going further

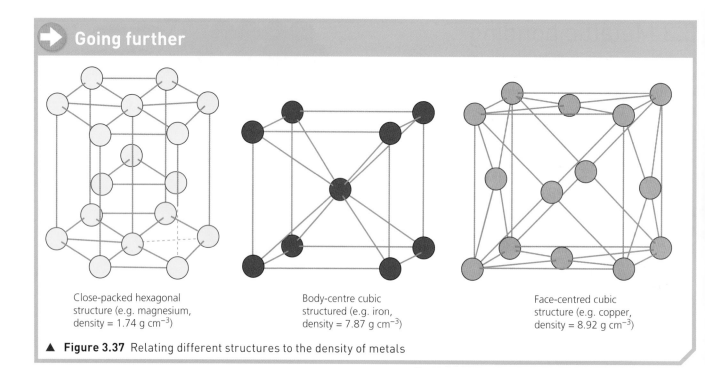

Close-packed hexagonal structure (e.g. magnesium, density = 1.74 g cm⁻³)

Body-centre cubic structured (e.g. iron, density = 7.87 g cm⁻³)

Face-centred cubic structure (e.g. copper, density = 8.92 g cm⁻³)

▲ **Figure 3.37** Relating different structures to the density of metals

Revision checklist

After studying Chapter 3 you should be able to:
✔ Explain how compounds form between metals and non-metals.
✔ Describe the properties of ionic compounds.
✔ Explain in terms of structure and bonding the properties of ionic compounds.
✔ Work out the formulae of ionic substances from the charges on the ions.
✔ Define redox reactions as reactions involving gain and loss of oxygen or as reactions involving gain and loss of electrons.
✔ Identify redox reactions as reactions involving gain and loss of electrons.
✔ Identify redox reactions by changes in oxidation number.
✔ Define redox reactions as involving simultaneous oxidation and reduction.
✔ Define oxidation as an increase in oxidation number, and reduction as a decrease in oxidation number.
✔ Identify redox reactions by the colour changes involved when using acidified potassium manganate(VII) or potassium iodide.
✔ Define an oxidising agent as a substance that oxidises another substance and is itself reduced during a redox reaction.

✔ Define a reducing agent as a substance that reduces another substance and is itself oxidised during a redox reaction.
✔ Identify oxidation and reduction in redox reactions.
✔ Identify oxidising agents and reducing agents in redox reactions.
✔ Explain how simple compounds form between non-metals by single covalent bonds.
✔ Explain how simple compounds containing multiple covalent bonds form between non-metals.
✔ Describe and explain the properties of simple covalent compounds.
✔ Describe the giant covalent structures of graphite and diamond.
✔ Describe the giant covalent structure of silicon(IV) oxide, SiO_2.
✔ Relate the structures of graphite and diamond to their uses.
✔ Describe the similarity in properties between diamond and silicon(IV) oxide, related to their structures.
✔ Describe the arrangement of particles in a solid metal.
✔ Recognise that the particles in a metal are held together by metallic bonds.
✔ Describe metallic bonding.
✔ Use metallic bonding to explain some of the properties of metals.

Exam-style questions

1 Define the following:
   a positive (+) ion [2]
   b negative (−) ion [2]
   c ionic bond [2]
   d electrostatic force of attraction. [2]

2 Using Table 3.1 write down the names of:
   a three atoms which would form an ion
      with a charge of +2 [3]
   b three atoms which would form an ion
      with a charge of +1 [3]
   c three atoms/groups of atoms which would
      form an ion with a charge of −2. [3]

3 Sketch diagrams to show the bonding in each
   of the following compounds:
   a calcium fluoride (CaF_2) [4]
   b oxygen (O_2) [4]
   c magnesium chloride ($MgCl_2$) [4]
   d tetrachloromethane (CCl_4). [4]

4 Use the information given in Table 3.1
   on p. 35 to deduce the formula for:
   a silver oxide
   b zinc chloride
   c potassium sulfate
   d calcium nitrate
   e iron(II) nitrate
   f copper(II) carbonate
   g iron(III) hydroxide
   h aluminium fluoride. [8]

5 Atoms of elements X, Y and Z have 16, 17 and
   19 electrons, respectively. Atoms of argon have
   18 electrons.
   a Determine the formulae of the compounds
      formed by the combination of the atoms of
      the elements:
      i X and Z [1]
      ii Y and Z [1]
      iii X with itself. [1]
   b In each of the cases shown in **a i–a iii** above,
      identify the type of chemical bond formed. [3]
   c Give two properties you would expect each
      of the compounds formed in **a ii** and **a iii**
      to have. [4]

6 Explain the following statements.
   a Ammonia is a gas at room temperature. [3]
   b The melting points of sodium chloride
      and iodine are very different. [8]
   c Metals generally are good conductors
      of electricity. [3]
   d Graphite acts as a lubricant but diamond
      does not. [3]

4 Stoichiometry – chemical calculations

FOCUS POINTS

★ How can I calculate the mass of products in a reaction?
★ What is relative atomic mass?
★ How is a mole and the Avogadro constant useful in balancing calculations?
★ How can I determine the formula of a compound?

In this chapter you will learn how to calculate the masses of products formed in a chemical reaction, when told the quantity of reactants you start with, and also how to calculate the mass of starting materials you need to produce a certain mass of product. These are the type of calculations your teacher does before they write the methods for student practicals. Have you ever thought how your teachers were able to get it right every time?

We will look at the amounts of substances used and formed during reactions involving solids, gases and solutions. You need to know whether you actually got the right amount of product in your experiment, and we will also look at how well a reaction went.

By the end of this chapter we hope that you will be confident in using numbers and balanced chemical reactions to help you to discover these things.

4.1 Relative atomic mass

There are at present 118 different elements known. The atoms of these elements differ in mass because of the different numbers of protons, neutrons and electrons they contain. The actual mass of one atom is very small. For example, the mass of a single atom of sulfur is around:

0.000 000 000 000 000 000 000 053 16 g

Such small quantities are not easy to work with and, as you saw in Chapter 3, a scale called the **relative atomic mass**, A_r, scale is used. In this scale an atom of carbon is given a relative atomic mass, A_r, of 12.00. All other atoms of the other elements are given a relative atomic mass compared to that of carbon.

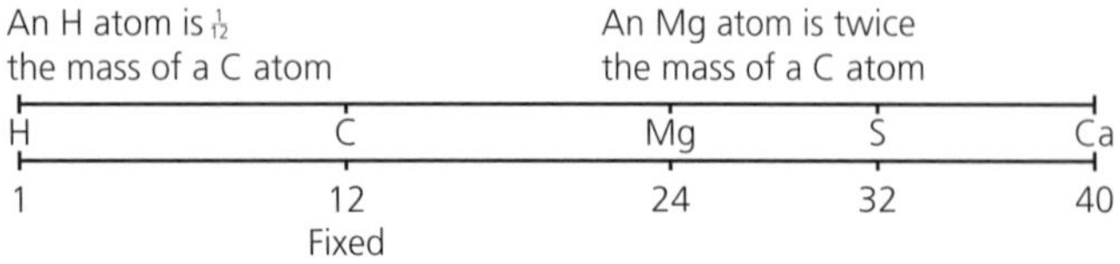

An H atom is $\frac{1}{12}$ the mass of a C atom

An Mg atom is twice the mass of a C atom

H	C	Mg	S	Ca
1	12	24	32	40
	Fixed			

▲ **Figure 4.1** The relative atomic masses of the elements H, C, Mg, S and Ca

> **Key definition**
> **Relative atomic mass, A_r,** is the average mass of the isotopes of an element compared to 1/12th of the mass of an atom of ^{12}C.

Reacting masses

Chemists often need to be able to show the relative masses of the atoms involved in a chemical process.

> ### ❓ Worked example
>
> What mass of carbon dioxide would be produced if 6 g of carbon was completely burned in oxygen gas?
>
> $$C + O_2 \rightarrow CO_2$$
>
> Instead of using the actual masses of atoms we use the relative atomic mass to help us answer this type of question.
>
> In this example we can work out the **relative molecular mass, M_r,** of molecules such as O_2 and CO_2 using the relative atomic masses of the atoms they are made from. The relative molecular mass is the sum of the relative atomic masses of all those elements shown in the molecular formula of the substance. The molecular formula of oxygen is O_2 and it shows that one molecule of oxygen contains two oxygen atoms. Each oxygen atom has a relative atomic mass of 16, so O_2 has a relative molecular mass of $2 \times 16 = 32$.

The molecular formula of carbon dioxide is CO_2 and it shows that in one molecule of carbon dioxide there is one carbon atom and two oxygen atoms. The carbon atom has a relative atomic mass of 12 and each oxygen atom has a relative atomic mass of 16, so CO_2 has a relative molecular mass of $12 + (2 \times 16) = 44$.

> **Key definitions**
>
> **Relative molecular mass**, M_r, is the sum of the relative atomic masses. **Relative formula mass**, M_r, is used for ionic compounds.

So we can now use the equation to answer the question: What mass of carbon dioxide would be produced if 6 g of carbon was completely combusted?

$$C + O_2 \rightarrow CO_2$$
$$12 \quad 32 \quad\quad 44$$

We can convert these relative masses to actual masses by adding mass units, g, which gives:

$$C + O_2 \rightarrow CO_2$$
$$12 \quad 32 \quad\quad 44$$
$$12\,g \quad 32\,g \quad\quad 44\,g$$

The above calculation shows that if 12 g of carbon were burned completely then 44 g of carbon dioxide gas would be formed. So, 6 g of carbon burning would result in the formation of 22 g of carbon dioxide gas.

In this example, we have assumed that the carbon is completely combusted. This is only possible because there is enough oxygen in the air to react with all the carbon. The reactant which is not completely used up in a reaction is referred to as the reactant in excess. Once the reaction is complete, some of this reactant will be left over.

> **Test yourself**
>
> 1 What mass of carbon dioxide gas would be produced if 10 g of calcium carbonate reacted with an excess of hydrochloric acid?
> 2 What mass of sulfur dioxide would be produced if 64 tonnes of sulfur was completely reacted with oxygen gas?

Chemists often need to know how much of a substance has been formed or used up during a chemical reaction. This is particularly important in the chemical industry, where the substances being reacted (the reactants) and the substances being produced (the products) may be worth a great deal of money. Waste costs money!

To solve this problem, the chemical industry needs a way of counting atoms, ions or molecules. As atoms, ions and molecules are very tiny particles, it is impossible to measure out a dozen or even a hundred of them. Instead, chemists weigh out a very large number of particles.

This number is 6.02×10^{23} atoms, ions or molecules and is called **Avogadro's constant** after the famous Italian scientist Amedeo Avogadro (1776–1856). An amount of substance containing 6.02×10^{23} particles is called a **mole** (often abbreviated to mol).

> **Key definition**
>
> The **mole**, symbol mol, is the unit of amount of substance. One mole contains 6.02×10^{23} particles, e.g. atoms, ions, molecules. This number is called the Avogadro constant.

4.2 Calculating moles

In Chapter 3 we looked at how we can compare the masses of all the other atoms with the mass of carbon atoms. This is the basis of the relative atomic mass scale. Chemists have found by experiment that if you take the relative atomic mass of an element in grams, it always contains 6.02×10^{23} or 1 mole of its atoms.

Moles and elements

For the elements we can see that, for example, the relative atomic mass (A_r) of iron is 56, so 1 mole of iron is 56 g. Therefore, 56 g of iron contains 6.02×10^{23} atoms.

The A_r for aluminium is 27. In 27 g of aluminium it is found that there are 6.02×10^{23} atoms. Therefore, 27 g of aluminium is 1 mole of aluminium atoms.

So, we can calculate the mass of a substance present in any number of moles using the relationship:

$$\text{mass (in grams)} = \text{number of moles} \times \text{molar mass of the element}$$

The molar mass of an element or compound is the mass of 1 mole of the element or compound and has units of g/mol.

So, a mole of the element magnesium is 6.02×10^{23} atoms of magnesium and a mole of the element carbon is 6.02×10^{23} atoms of carbon (Figure 4.2).

a A mole of magnesium

b A mole of carbon

▲ **Figure 4.2**

? ## Worked example

Calculate the number of atoms of carbon in:
a   0.5 moles
b   0.1 moles

Molar mass of carbon contains 6.02×10^{23} atoms of carbon so 0.5 moles would contain half this amount:

Atoms of carbon in 0.5 moles $= 6.02 \times 10^{23} \times 0.5 = 3.01 \times 10^{23}$ atoms

In a similar way:

Atoms of carbon in 0.1 moles $= 6.02 \times 10^{23} \times 0.1 = 6.02 \times 10^{22}$ atoms

---

? ## Worked example

What mass of hydrogen gas would be produced if 46 g of sodium completely reacted with water?

First write down the balanced chemical equation:

$$2Na + 2H_2O \rightarrow 2NaOH + H_2$$

Next find the relative atomic mass of sodium (from the Periodic Table (p. 130)) and work out the relative formula masses of water, sodium hydroxide and hydrogen gas. The term **relative formula mass** is used when the compound is ionic, in this case sodium hydroxide. It is used and calculated in the same way as we have used relative molecular mass for covalently bonded compounds, such as water and hydrogen.

Relative atomic mass of sodium is 23.

Relative molecular mass of water, H_2O, is $(2 \times 1) + 16 = 18$

Relative formula mass of sodium hydroxide is $23 + 16 + 1 = 40$

Relative molecular mass of hydrogen gas, H_2, is $2 \times 1 = 2$

Now write these masses under the balanced chemical equation taking into account the numbers used to balance the equation.

$$2Na \quad + \quad 2H_2O \quad \rightarrow \quad 2NaOH \quad + \quad H_2$$
$$2 \times 23 = 46 \quad 2 \times 18 = 36 \quad 2 \times 40 = 80 \quad 2$$

These relative masses can now be converted into actual or reacting masses by putting in mass units, for example, grams.

$$2Na \quad + \quad 2H_2O \quad \rightarrow \quad 2NaOH \quad + \quad H_2$$
$$2 \times 23 = 46 \quad 2 \times 18 = 36 \quad 2 \times 40 = 80 \quad 2$$
$$46\,g \qquad\qquad 36\,g \quad \rightarrow \quad 18\,g \qquad\qquad 2\,g$$

So the answer to the question of what mass of hydrogen would be produced if 46 g of sodium was reacted with water is 2 g.

Worked example

Calculate the mass of **a** 2 moles of iron and **b** 0.25 mole of iron. (A_r: Fe = 56)

a   mass of 2 moles of iron
= number of moles × relative atomic mass (A_r)
= 2 × 56
= 112 g

b   mass of 0.25 mole of iron
= number of moles × relative atomic mass (A_r)
= 0.25 × 56
= 14 g

If we know the mass of the element, then it is possible to calculate the number of moles of that element using:

$$\text{number of moles} = \frac{\text{mass of the element}}{\text{molar mass}}$$

 Worked example

Calculate the number of moles of aluminium present in **a** 108 g and **b** 13.5 g of the element. (A_r: Al = 27)

a   number of moles of aluminium

$$= \frac{\text{mass of aluminium}}{\text{molar mass of aluminium}}$$

$$= \frac{108}{27}$$

= 4 moles

b   number of moles of aluminium

$$= \frac{\text{mass of aluminium}}{\text{molar mass of aluminium}}$$

$$= \frac{13.5}{27}$$

= 0.5 mole

4.3 Moles and compounds

The idea of the mole can also be used with compounds (Figure 4.3).

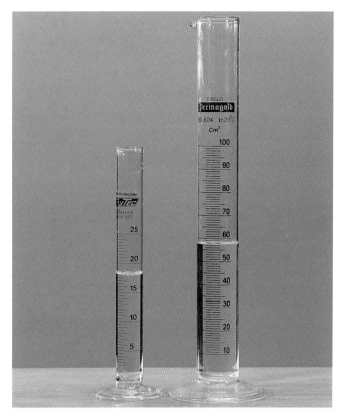

▲ **Figure 4.3** One mole of water (H_2O) (left) and 1 mole of ethanol (C_2H_5OH) (right) in separate measuring cylinders

For example, consider the molar mass of water (H_2O) molecules (A_r: H = 1; O = 16).

From the formula of water, H_2O, you will see that 1 mole of water molecules contains 2 moles of hydrogen (H) atoms and 1 mole of oxygen (O) atoms. The molar mass of water molecules is therefore:

$$(2 \times 1) + (1 \times 16) = 18 \text{ g}$$

The mass of 1 mole of a compound is called its **molar mass**; it has units of g/mol. If you write the molar mass of a compound without any units then it is the relative molecular mass (M_r). So the relative molecular mass of water is 18.

Now follow these examples to help you understand more about moles and compounds.

❓ Worked example

What is **a** the molar mass and **b** the relative molecular mass, M_r, of ethanol, C_2H_5OH? (A_r: H = 1; C = 12; O = 16)

a One mole of C_2H_5OH contains 2 moles of carbon atoms, 6 moles of hydrogen atoms and 1 mole of oxygen atoms. Therefore:

molar mass of ethanol = $(2 \times 12) + (6 \times 1) + (1 \times 16)$
$$= 46 \text{ g/mol}$$

b The relative molecular mass of ethanol is 46.

❓ Worked example

What is **a** the molar mass and **b** the relative molecular mass of nitrogen gas, N_2? (A_r: N = 14)

a Nitrogen is a diatomic gas. Each nitrogen molecule contains two atoms of nitrogen. Therefore:

molar mass of $N_2 = 2 \times 14$
$$= 28 \text{ g/mol}$$

b The relative molecular mass of N_2 is 28.

The mass of a compound found in any number of moles can be calculated using the relationship:

$$\text{mass of compound} = \text{number of moles of the compound} \times \text{molar mass of the compound}$$

❓ Worked example

Calculate the mass of **a** 3 moles and **b** 0.2 moles of carbon dioxide gas, CO_2. (A_r: C = 12; O = 16)

a One mole of CO_2 contains 1 mole of carbon atoms and 2 moles of oxygen atoms. Therefore:

molar mass of $CO_2 = (1 \times 12) + (2 \times 16)$
$$= 44 \text{ g/mol}$$

mass of 3 moles of CO_2 = number of moles × molar mass of CO_2
$$= 3 \times 44$$
$$= 132 \text{ g}$$

b mass of 0.2 mole of CO_2 = number of moles × molar mass of CO_2
$$= 0.2 \times 44$$
$$= 8.8 \text{ g}$$

If we know the mass of the compound, then we can calculate the number of moles of the compound using the relationship:

$$\text{number of moles of compound} = \frac{\text{mass of compound}}{\text{molar mass of the compound}}$$

❓ Worked example

Calculate the number of moles of magnesium oxide, MgO, in **a** 80 g and **b** 10 g of the compound. (A_r: O = 16; Mg = 24)

a One mole of MgO contains 1 mole of magnesium atoms and 1 mole of oxygen atoms. Therefore:

molar mass of MgO = $(1 \times 24) + (1 \times 16) = 40$ g/mol
number of moles of MgO in 80 g

$$= \frac{\text{mass of MgO}}{\text{molar mass of MgO}} = \frac{80}{40}$$

$$= 2 \text{ moles}$$

b number of moles of MgO in 10 g

$$= \frac{\text{mass of MgO}}{\text{molar mass of MgO}} = \frac{10}{40}$$

$$= 0.25 \text{ mole}$$

Moles and gases

Many substances exist as gases. If we want to find the number of moles of a gas we can do this by measuring the volume rather than the mass.

Chemists have shown by experiment that one mole of any gas occupies a volume of approximately 24 dm³ (24 litres) at room temperature and pressure (r.t.p.). This quantity is also known as the **molar gas volume.**

Therefore, it is relatively easy to convert volumes of gases into moles and moles of gases into volumes using the following relationship:

$$\text{number of moles of a gas} = \frac{\text{volume of the gas (in dm}^3 \text{at r.t.p.)}}{24 \text{ dm}^3}$$

or

$$\text{volume of a gas (in dm}^3 \text{ at r.t.p.)} = \text{number of moles of gas} \times 24 \text{ dm}^3$$

We will use these two relationships to help us answer some questions concerning gases.

❓ Worked example

Calculate the number of moles of ammonia gas, NH_3, in a volume of 72 dm³ of the gas measured at r.t.p.

$$\text{number of moles of ammonia} = \frac{\text{volume of ammonia in dm}^3}{24 \text{ dm}^3}$$

$$= \frac{72}{24} = 3$$

Worked example

Calculate the volume of carbon dioxide gas, CO_2, occupied by
a 5 moles and **b** 0.5 mole of the gas measured at r.t.p.

a   volume of CO_2 = number of moles of CO_2 × 24 dm^3
                = 5 × 24 = 120 dm^3

b   volume of CO_2 = number of moles of CO_2 × 24 dm^3
                = 0.5 × 24 = 12 dm^3

The volume occupied by 1 mole of any gas must contain $6.02 × 10^{23}$ molecules. Therefore, it follows that equal volumes of all gases measured at the same temperature and pressure must contain the same number of molecules. This hypothesis was also first put forward by Amedeo Avogadro and is called **Avogadro's Law.**

Moles and solutions

Chemists often need to know the concentration of a solution. Sometimes **concentration** is measured in grams per cubic decimetre (g/dm^3), but more often concentration is measured in moles per cubic decimetre (mol/dm^3).

> **Key definition**
> **Concentration** can be measured in g/dm^3 or mol/dm^3.

When 1 mole of a substance is dissolved in water and the solution is made up to 1 dm^3 (1000 cm^3), a 1 molar (1 mol/dm^3) solution is produced. Chemists do not always need to make up such large volumes of solution.

A simple method of calculating the concentration uses the relationship:

$$\text{concentration (in } mol/dm^3) = \frac{\text{number of moles}}{\text{volume (in } dm^3)}$$

It is very easy to change a volume given in cm^3 into one in dm^3 by simply dividing the volume in cm^3 by 1000. For example, 250 cm^3 is 0.25 dm^3.

 Worked example

Calculate the concentration (in mol/dm^3) of a solution of sodium hydroxide, NaOH, which was made by dissolving 10 g of solid sodium hydroxide in water and making up to 250 cm^3. (A_r: H = 1; O = 16; Na = 23)

1 mole of NaOH contains 1 mole of sodium, 1 mole of oxygen and 1 mole of hydrogen. Therefore:

$$\text{molar mass of NaOH} = (1 × 23) + (1 × 16) + (1 × 1)$$
$$= 40 \text{ g}$$

$$\frac{\text{number of moles}}{\text{NaOH in 10 g}} = \frac{\text{mass of NaOH}}{\text{molar mass of NaOH}}$$

$$= \frac{10}{40} = 0.25$$

$$\left(250 \text{ } cm^3 = \frac{250}{1000} \text{ } dm^3 = 0.25 \text{ } dm^3\right)$$

$$\frac{\text{concentration of the}}{\text{NaOH solution}} = \frac{\text{number of moles of NaOH}}{\text{volume of solution (} dm^3)}$$

$$= \frac{0.25}{0.25} = 1 \text{ } mol/dm^3$$

Sometimes chemists need to know the mass of a substance that has to be dissolved to prepare a known volume of solution at a given concentration. A simple method of calculating the number of moles, and so the mass of substance needed, is by using the relationship:

$$\frac{\text{number of}}{\text{moles}} = \frac{\text{concentration}}{\text{(in } mol/dm^3)} × \frac{\text{volume in solution}}{\text{(in } dm^3)}$$

 Worked example

Calculate the mass of potassium hydroxide, KOH, that needs to be used to prepare 500 cm^3 of a 2 mol/dm^3 solution in water. (A_r: H = 1; O = 16; K = 39)

number of moles of KOH

$$= \frac{\text{concentration of solution}}{\text{(} mol/dm^3)} × \frac{\text{volume of solution}}{\text{(} dm^3)}$$

$$= 2 × \frac{500}{1000} = 1$$

1 mole of KOH contains 1 mole of potassium, 1 mole of oxygen and 1 mole of hydrogen. Therefore:

molar mass of KOH

$$= (1 × 39) + (1 × 16) + (1 × 1)$$
$$= 56 \text{ g}$$

Therefore:

    mass of KOH in 1 mole = 56 g

Test yourself

Use these values of A_r to answer the questions below.
H = 1; C = 12; N = 14; O = 16; Ne = 20; Na = 23; Mg = 24;
S = 32; Cl = 35.5; K = 39; Fe = 56; Cu = 63.5; Zn = 65.
One mole of any gas at r.t.p. occupies 24 dm^3.

3 Calculate the number of moles in:
   a 2 g of neon atoms
   b 4 g of magnesium atoms
   c 24 g of carbon atoms.
4 Calculate the number of atoms in 1 mole of:
   a calcium
   b carbon dioxide
   c methane.
5 Calculate the mass of:
   a 0.1 mol of oxygen molecules
   b 5 mol of sulfur atoms
   c 0.25 mol of sodium atoms.
6 Calculate the number of moles in:
   a 9.8 g of sulfuric acid (H_2SO_4)
   b 40 g of sodium hydroxide (NaOH)
   c 720 g of iron(II) oxide (FeO).
7 Calculate the mass of:
   a 2 mol of zinc oxide (ZnO)
   b 0.25 mol of hydrogen sulfide (H_2S)
   c 0.35 mol of copper(II) sulfate ($CuSO_4$).
8 Calculate the number of moles at r.t.p. in:
   a 2 dm^3 of carbon dioxide (CO_2)
   b 240 dm^3 of sulfur dioxide (SO_2)
   c 20 cm^3 of carbon monoxide (CO).
9 Calculate the volume of:
   a 0.3 mol of hydrogen chloride (HCl)
   b 4.4 g of carbon dioxide
   c 34 g of ammonia (NH_3).
10 Calculate the concentration of solutions containing:
   a 0.2 mol of sodium hydroxide dissolved in water and made up to 100 cm^3
   b 9.8 g of sulfuric acid dissolved in water and made up to 500 cm^3.
11 Calculate the mass of:
   a copper(II) sulfate ($CuSO_4$) which needs to be used to prepare 500 cm^3 of a 0.1 mol/dm^3 solution
   b potassium nitrate (KNO_3) which needs to be used to prepare 200 cm^3 of a 2 mol/dm^3 solution.

4.4 Calculating formulae

If we have 1 mole of a compound, then the formula shows the number of moles of each element in that compound. For example, the formula for lead(II)

bromide is $PbBr_2$. This means that 1 mole of lead(II) bromide contains 1 mole of lead ions and 2 moles of bromide ions. If we do not know the formula of a compound, we can find the masses of the elements present experimentally, and we can then use these masses to work out the formula of that compound.

Practical skills

Experiment to find the formula of magnesium oxide

Safety
● Eye protection must be worn.
● Take care when handling hot apparatus.

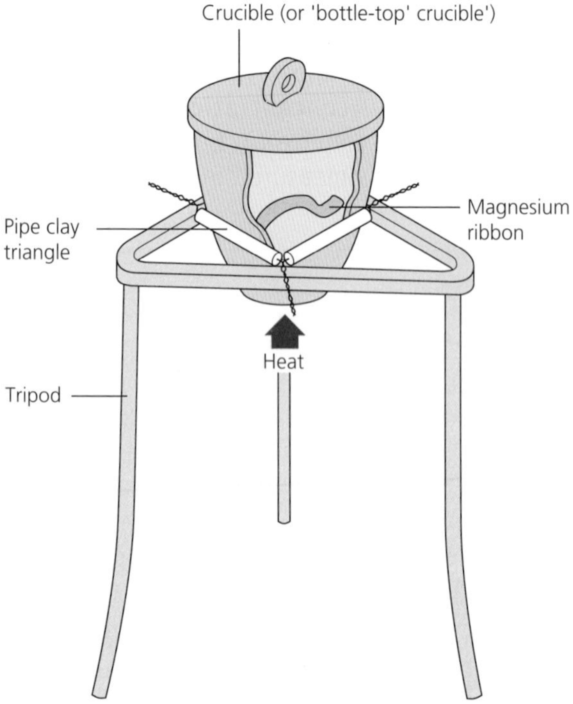

▲ **Figure 4.4** Apparatus used to determine magnesium oxide's formula

If a known mass of magnesium ribbon is heated strongly in a crucible (or 'bottle-top' crucible) with a lid, it reacts with oxygen to form magnesium oxide.

magnesium + oxygen → magnesium oxide

To allow oxygen into the crucible the lid needs to be lifted briefly during the heating process.

To heat to constant mass:

- Heat strongly for 5 minutes, then the crucible and contents should be allowed to cool and reweighed.
- The above step should be repeated until a constant mass is obtained.

When the magnesium is heated it can be seen that:

- the shiny magnesium metal burns brightly to form a white powder, magnesium oxide
- the powder is very fine and when the lid is lifted it can be seen to rise out of the crucible (which is the reason the lid is lifted only briefly).

▼ **Table 4.1** Data from the magnesium oxide experiment

Mass of crucible	14.63 g
Mass of crucible and magnesium	14.87 g
Mass of crucible and magnesium oxide	15.03 g

From these data we can find the formula of magnesium oxide by following these steps:

1 Calculate the mass of magnesium metal used in this experiment.
2 Calculate the mass of oxygen which reacts with the magnesium.
3 Calculate the number of moles of magnesium used.
4 Calculate the number of moles of oxygen atoms which react with the magnesium.
5 Determine the simplest ratio of moles of magnesium to moles of oxygen.
6 Determine the simplest formula for magnesium oxide.

This formula is the **empirical formula** of the compound.

> **Key definition**
>
> The **empirical formula** of a compound is the simplest whole number ratio of the different atoms or ions in a compound.

7 Why is it important to lift the lid briefly while heating the crucible?
8 What are the main sources of error in this experiment?

? Worked example

In an experiment an unknown organic compound was found to contain 0.12 g of carbon and 0.02 g of hydrogen. Calculate the empirical formula of the compound. (A_r: H = 1; C = 12)

	C	H
Masses (g)	0.12	0.02
Number of moles	$\frac{0.12}{12} = 0.01$	$\frac{0.02}{1} = 0.02$
Ratio of moles	1	2
Empirical formula	CH$_2$	

From our knowledge of covalent bonding (Chapter 3, p. 37) we know that a molecule of this formula cannot exist. However, molecules with the following formulae do exist: C_2H_4, C_3H_6, C_4H_8 and C_5H_{10}. All of these formulae show the same ratio of carbon atoms to hydrogen atoms, CH_2, as our unknown. To find out which of these formulae is the actual formula for the unknown organic compound, we need to know the molar mass of the compound.

Using a mass spectrometer, the relative molecular mass (M_r) of this organic compound was found to be 56. We need to find out the number of empirical formulae units present:

M_r of the empirical formula unit

$= (1 \times 12) + (2 \times 1)$

$= 14$

Number of empirical formula units present

$= \dfrac{M_r \text{ of compound}}{M_r \text{ of empirical formula unit}} = \dfrac{56}{14}$

$= 4$

Therefore, the actual formula of the unknown organic compound is $4 \times CH_2 = C_4H_8$.

This substance is called butene. C_4H_8 is the **molecular formula** for this substance and shows the actual numbers of atoms of each element present in one molecule of the substance.

Sometimes the composition of a compound is given as a percentage by mass of the elements present. In cases such as this the procedure shown in the next example is followed.

? Worked example

Calculate the empirical formula of an organic compound containing 92.3% carbon and 7.7% hydrogen by mass. The M_r of the organic compound is 78. What is its molecular formula? (A_r: H = 1; C = 12)

	C	H
% by mass	92.3	7.7
Masses in 100 g	92.3 g	7.7 g
Number of moles	$\frac{92.3}{12} = 7.7$	$\frac{7.7}{1} = 7.7$
Ratio of moles	1	1
Empirical formula	CH	

M_r of the empirical formula unit CH

$= 12 + 1$

$= 13$

Number of empirical formula units present

$$= \frac{M_r \text{ of compound}}{M_r \text{ of empirical formula unit}} = \frac{78}{13}$$

$= 6$

The molecular formula of the organic compound is $6 \times CH = C_6H_6$. This is a substance called benzene.

Test yourself

Use the following values of A_r to answer the questions below: H = 1; C = 12; O = 16; Ca = 40.

12 Determine the empirical formula of an oxide of calcium formed when 0.4 g of calcium reacts with 0.16 g of oxygen.

13 Determine the empirical formula of an organic hydrocarbon compound which contains 80% by mass of carbon and 20% by mass of hydrogen. If the M_r of the compound is 30, what is its molecular formula?

4.5 Moles and chemical equations

When we write a balanced chemical equation, we are indicating the numbers of moles of reactants and products involved in the chemical reaction. Consider the reaction between magnesium and oxygen.

magnesium + oxygen → magnesium oxide

$$2Mg(s) + O_2(g) \rightarrow 2MgO(s)$$

This shows that 2 moles of magnesium react with 1 mole of oxygen to give 2 moles of magnesium oxide.

Using the ideas of moles and masses we can use this information to calculate the quantities of the different chemicals involved.

$2Mg(s)$	+	$O_2(g)$	→	$2MgO(s)$
2 moles		1 mole		2 moles
2×24		$1 \times (16 \times 2)$		$2 \times (24 + 16)$
= 48 g		= 32 g		= 80 g

You will notice that the total mass of reactants is equal to the total mass of product. This is true for any chemical reaction and it is known as the **Law of conservation of mass.** This law was understood by the Greeks but was first clearly formulated by Antoine Lavoisier in 1774. Chemists can use this idea to calculate masses of products formed and reactants used in chemical processes before they are carried out.

Solids

❓ Worked example

Lime (calcium oxide, CaO) is used in the manufacture of lime mortar, a mixture of lime, sand, aggregate and water. It is used as a binding material when building with brick. Lime is manufactured in large quantities in Pakistan (Figure 4.5) by heating limestone (calcium carbonate, $CaCO_3$).

▲ **Figure 4.5** At this factory in Pakistan, limestone is converted into lime and the lime is used to make cement

The equation for the process is:

$$CaCO_3(s) \rightarrow CaO(s) + CO_2(g)$$

1 mole	1 mole	1 mole
$40 + 12 + (3 \times 16)$	$40 + 16$	$12 + (2 \times 16)$
$= 100$ g	$= 56$ g	$= 44$ g

Calculate the amount of lime produced when 10 tonnes of limestone are heated (Figure 4.6). (A_r: C = 12; O = 16; Ca = 40)

$$1 \text{ tonne (t)} = 1000 \text{ kg}$$

$$1 \text{ kg} = 1000 \text{ g}$$

From this relationship between grams and tonnes we can replace the masses in grams by masses in tonnes:

$$CaCO_3(s) \rightarrow CaO(s) + CO_2(g)$$

100 t	56 t	44 t
10 t	5.6 t	14.4 t

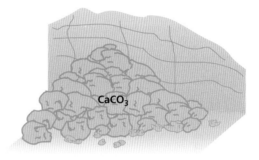

10 tonnes of limestone

Heat

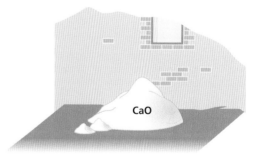

? tonnes of lime

▲ **Figure 4.6** How much lime is produced?

The equation now shows that 100 t of limestone will produce 56 t of lime. Therefore, 10 t of limestone will produce 5.6 t of lime.

Gases

Many chemical processes involve gases. The volume of a gas is measured more easily than its mass. This example shows how chemists work out the volumes of gaseous reactants and products needed using Avogadro's Law and the idea of moles.

? Worked example

Some rockets use hydrogen gas as a fuel. When hydrogen burns in oxygen it forms water vapour. Calculate the volumes of **a** $O_2(g)$ used and **b** water, $H_2O(g)$, produced if 960 dm³ of hydrogen gas, $H_2(g)$, were burned in oxygen (A_r: H = 1; O = 16). Assume 1 mole of any gas occupies a volume of 24 dm³.

$2H_2(g)$	+	$O_2(g)$	$\rightarrow$	$2H_2O(g)$
2 moles		1 mole		2 moles
2 × 24		1 × 24		2 × 24
= 48 dm³		= 24 dm³		= 48 dm³

Therefore:

(×2)	96 dm³	48 dm³	96 dm³
(×10)	960 dm³	480 dm³	960 dm³

When 960 dm³ of hydrogen are burned in oxygen:
a   480 dm³ of oxygen are required and
b   960 dm³ of $H_2O(g)$ are produced.

Solutions

? Worked example

Chemists usually carry out reactions using solutions. If they know the concentration of the solution(s) they are using, they can find out the quantities reacting.

Calculate the volume of 1 mol/dm³ solution of H_2SO_4 required to react completely with 6 g of magnesium. (A_r: Mg = 24).

number of moles of magnesium

$$= \frac{\text{mass of magnesium}}{\text{molar mass of magnesium}} = \frac{6}{24}$$

$$= 0.25$$

Mg(s)	+	$H_2SO_4(aq)$	$\rightarrow$	$MgSO_4(aq)$	+	$H_2(g)$
1 mole		1 mole		1 mole		1 mole
0.25 mol		0.25 mol		0.25 mol		0.25 mol

So 0.25 mol of $H_2SO_4(aq)$ is required. Using:

volume of $H_2SO_4(aq)$ (dm³)

$$= \frac{\text{moles of } H_2SO_4}{\text{concentration of } H_2SO_4 \text{ (mol/dm}^3)} = \frac{0.25}{1}$$

$$= 0.25 \text{ dm}^3 \text{ or } 250 \text{ cm}^3$$

? Worked example

40 cm³ of 0.2 mol/dm³ solution of hydrochloric acid just neutralised 20 cm³ of sodium hydroxide solution in a titration (Chapter 8, p. 124).

What is the concentration of sodium hydroxide solution in g/dm³ used in this neutralisation reaction?

number of moles of HCl used

= concentration (mol/dm³) × volume (dm³) = 0.2 × 0.04

= 0.008

HCl(aq)	+	NaOH(aq)	$\rightarrow$	NaCl(aq)	+	$H_2O(l)$
1 mole		1 mole		1 mole		1 mole
0.008 mol		0.008 mol		0.008 mol		0.008 mol

You will see that 0.008 mole of NaOH was present. The concentration of the NaOH(aq) is given by:

concentration of NaOH (mol/dm³)

$$= \frac{\text{number of moles of NaOH}}{\text{volume of NaOH (dm}^3)} = \frac{0.008}{0.02}$$

$$\left(\text{volume of NaOH in dm}^3 = \frac{20}{1000} = 0.02\right)$$

$$= 0.4 \text{ mol/dm}^3$$

Now we have the concentration in mol/dm³ we can easily convert this in g/dm³ using the relationship below:

Concentration of a solution in g/dm³	=	concentration in mol/dm³	×	molar mass of the substance

So the concentration of the NaOH in g/dm³ = 0.4 × (23 + 16 + 1)
= 16 g/dm³

Percentage yield

Chemical reactions rarely produce the predicted amount of product from the masses of reactants in the reaction as they are not 100% efficient.

An example of this is the reaction of carbon with oxygen to produce carbon dioxide gas.

$$C(s) + O_2(g) \rightarrow CO_2(g)$$

The equation for the reaction states that 1 mole of carbon reacts with oxygen to give 1 mole of carbon dioxide gas.

If you burn 12 g, 1 mole, of carbon to make CO_2, then the amount of carbon dioxide expected is 44 g, 1 mole of CO_2. The **theoretical yield** of carbon dioxide from this reaction is 44 g. This only occurs, however, if the reaction is 100% efficient.

In reality the mass of carbon dioxide you will get will be less than 44 g, because another reaction can also occur between carbon and oxygen. Some of the carbon reacts to make carbon monoxide, CO.

$$2C(s) + O_2(g) \rightarrow 2CO(g)$$

The **percentage yield** of the reaction is based on the amount of carbon dioxide that is actually produced against what should have been produced if the reaction was 100% efficient.

For example, if 12 g of carbon was burned in excess oxygen and only 28 g of carbon dioxide was produced, the percentage yield can be worked out:

$$\text{percentage yield of carbon dioxide} = \frac{28}{44} \times 100$$

$$= 63.6\%$$

In general:

$$\text{percentage yield} = \frac{\text{actual yield}}{\text{theoretical yield}} \times 100$$

Percentage composition

Percent composition is used to describe the percent by mass of each element in a compound. It is found by dividing the mass of a particular element in the compound by the molar mass and then multiplying by 100 to give a percentage.

For example, the percentage composition of magnesium oxide, MgO, can be found by using the calculations shown below.

The molar mass of MgO is 24 + 16 = 40 g. Of this 24 g is magnesium and 16 g is oxygen.

$$\% \, Mg = \frac{24}{40} \times 100 = 60\%$$

$$\% \, O = \frac{16}{40} \times 100 = 40\%$$

Percentage purity

In Chapter 2 (p. 14), we saw that the **purity** of a substance is very important. If a factory makes medicines or chemicals used in food then the purity of the product is crucial, as the impurities may harm the people using the medicine or eating the food.

$$\text{percentage purity} = \frac{\text{mass of the pure product}}{\text{mass of the impure product obtained}} \times 100\%$$

Sodium hydrogencarbonate, $NaHCO_3$, is used in the manufacture of some toothpastes and as a raising agent in food production. The purity of this substance can be obtained by measuring how much carbon dioxide is given off.

> **?** **Worked example**
>
> 84 g of sodium hydrogencarbonate was thermally decomposed and 11.5 dm³ of carbon dioxide gas was collected at room temperature and pressure (r.t.p.).
>
> The equation for the reaction is:
>
> $$2NaHCO_3(s) \rightarrow Na_2CO_3(s) + H_2O(g) + CO_2(g)$$
> $$\quad\; 2 \text{ moles} \qquad\quad 1 \text{ mole}$$
>
> **Step 1**: Calculate the relative formula mass of sodium hydrogencarbonate (A_r: Na = 23; C = 12; O = 16; H = 1)
>
> The relative formula mass of $NaHCO_3$ = 84
>
> **Step 2**: 2 moles of $NaHCO_3$ produces 1 mole of CO_2.
>
> 168 g of $NaHCO_3$ would give 44 g of CO_2, which would have a volume of 24 dm³ at r.t.p.
>
> 84 g of $NaHCO_3$ should give 12 dm³ of CO_2 at r.t.p.
>
> Hence the mass of $NaHCO_3$ in the sample was
>
> $$84 \times \frac{11.5}{12} = 80.5 \text{ g}$$
>
> **Step 3**: Calculate the percentage purity.
>
> There is 80.5 g of sodium hydrogencarbonate in the 84 g sample.
>
> $$\text{percentage purity} = \frac{80.5}{84} \times 100\% = 95.8\%$$

In a chemical reaction, not all the reactants are always completely used up. Usually one of the reactants will remain at the end of the reaction because it has nothing to react with. This is the excess reactant. The reactant which has been completely used up and does not remain at the end of the reaction is called the **limiting reactant**. The amount of product that can be obtained is determined by the limiting reactant. This is often the case when acids are used to make salts by their reactions with bases or carbonates (see Chapter 8, p. 117). An excess of the base or carbonate is used to ensure the acid has completely reacted. The acid would be the limiting reactant as it determines the amount of the salt produced.

In a chemical reaction it is important to be able to determine which reactant is the limiting reactant as this will allow you to find the maximum mass of products which can be produced.

 Worked example

How much copper(II) oxide could be formed if 16 g of copper reacted with 20 g of oxygen gas?

Step 1: Write down the balanced chemical equation for the reaction.

$$2Cu(s) + O_2(g) \rightarrow 2\ CuO(s)$$

2 moles   1 mole

Step 2: Convert the mass of each reactant into moles:

Moles of Cu used = 16/64 = 0.25 moles

Moles of O_2 used = 20/(2×16) = 0.625 moles

Step 3: Now look at the ratio of the moles in the balanced equation:

2 moles of Cu would react with 1 mole of O_2

0.25 moles of Cu would react with 0.25/2 = 0.125 moles of O_2

So Cu will be the limiting reactant, because it would be used up, but there would be 0.625 − 0.125 = 0.5 moles of O_2 remaining at the end of the reaction. The oxygen would have been used in excess.

Step 4: Now we know which is the limiting reactant we can use the mole ratio between it and the product to find the mass of copper(II) oxide we could produce.

0.25 moles of Cu would give 0.25 moles of CuO

Mass of CuO which could be produced = 0.25 × (64 + 16) = 20 g

14 Calculate the mass of sulfur dioxide produced by burning 16 g of sulfur in an excess of oxygen.
15 Calculate the mass of sulfur which, when burned in excess oxygen, produces 640 g of sulfur dioxide.
16 Calculate the mass of copper required to produce 159 g of copper(II) oxide when heated in excess oxygen.
17 A rocket uses hydrogen as a fuel. Calculate the volume of hydrogen used to produce 24 dm³ of water ($H_2O(g)$).
18 Calculate the volume of 2 mol/dm³ solution of sulfuric acid required to react with 24 g of magnesium.
19 20 cm³ of 0.2 mol/dm³ solution of hydrochloric acid just neutralised 15 cm³ of potassium hydroxide solution in a titration (see Chapter 8, p. 124). What is the concentration of potassium hydroxide solution used in this neutralisation reaction?
20 Calculate the percentage composition of ammonium nitrate, NH_4NO_3.
21 What is the maximum mass of magnesium oxide that could be produced if 48 g of magnesium reacted with 38 g of oxygen gas?

Revision checklist

After studying Chapter 4 you should be able to:
✔ Define and, given data, calculate the molecular and empirical formulae of a compound.
✔ Describe the terms relative atomic, molecular and formula mass.
✔ State that the mole is the unit of amount of substance and that it contains 6.02×10^{23} particles.

✔ Carry out calculations to find masses, moles, volumes and concentrations in reactions using solids gases and solutions.
✔ State that concentration has units of g/dm³ and mol/dm³ and convert between them.
✔ Calculate percentage yield and purity.
✔ Calculate the percentage composition of a compound.

Exam-style questions

Use the data in the table below to answer the questions.

Element	A_r
H	1
C	12
N	14
O	16
Na	23
Mg	24
Si	28
S	32
Cl	35.5
Fe	56

1 Calculate the mass of:
  a 1 mole of:
    i chlorine molecules
    ii iron(III) oxide.
  b 0.5 moles of:
    i magnesium nitrate
    ii ammonia.
2 Calculate the volume occupied, at r.t.p., by the following gases. (One mole of any gas occupies a volume of 24 dm³ at r.t.p.)
  a 12.5 moles of sulfur dioxide gas
  b 0.15 mole of nitrogen gas.

3 Calculate the number of moles of gas present in the following:
  a 36 cm³ of sulfur dioxide
  b 144 dm³ of hydrogen sulfide.
4 Use the following experimental information to determine the empirical formula of an oxide of silicon.

  Mass of crucible = 18.20 g

  Mass of crucible + silicon = 18.48 g

  Mass of crucible + oxide of silicon = 18.80 g

5 a Calculate the empirical formula of an organic liquid containing 26.67% of carbon and 2.22% of hydrogen, with the rest being oxygen.
  b The M_r of the liquid is 90. What is its molecular formula?
6 Iron is extracted from its ore, hematite, in a blast furnace. The main extraction reaction is:

$$Fe_2O_3(s) + 3CO(g) \rightarrow 2Fe(s) + 3CO_2(g)$$

  a Identify the reducing agent in this process.
  b Give the oxide of iron shown in the equation.
  c Explain why this is a redox reaction.
  d Calculate the mass of iron which will be produced from 640 tonnes of hematite.

Electrochemistry

5

FOCUS POINTS

★ What is the process of electrolysis?
★ How is electrolysis useful to us?
★ How do we make use of the Earth's aluminium?
★ Will fuel cells be used for transport in the future?

In this chapter you will look at the process of electrolysis and the industries based around the process of electrolysis. You will look at the electrolysis of various substances in the molten state, and in aqueous solution. You will see how useful electrolysis is across manufacturing and fuel production industries, by considering the extraction of aluminium, chlor-alkali industry and the industry based around electroplating. Finally, you will learn about fuel cells and how these cells may be used in our methods of transport.

5.1 Electricity and chemistry

Take a look at the photographs in Figure 5.1. What do you think they all have in common?

They all involve electricity and the process known as **electrolysis**. Electrolysis is the breakdown of an ionic compound, either molten or in solution, by the passage of electricity through it. The substance that is decomposed is called the **electrolyte** (Figure 5.2). An electrolyte is a substance that conducts electricity when in the molten state or in solution.

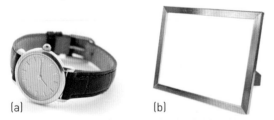

(a)   (b)

(c)

a This watch has a thin coating of gold over steel; the thin coating is produced by electrolysis
b This picture frame has been silver plated using an electroplating process involving electrolysis
c Aluminium is produced by electrolysis

▲ **Figure 5.1**

The electricity is carried through the electrolyte by **ions**. In the molten state and in aqueous or water solution, the ions are free to move to the appropriate electrodes due to weakened forces of attraction between them.

>> Substances that do not conduct electricity when in the molten state or in solution are called non-electrolytes.
>> Substances that conduct electricity to a small extent in the molten state or in solution are called weak electrolytes.

The electric current enters and leaves the electrolyte through **electrodes**, which are usually made of unreactive metals, such as platinum or the non-metal carbon (graphite). These are said to be **inert electrodes** because they do not react with the products of electrolysis. The two electrodes are the **cathode**, the negative electrode which attracts **cations** (positively charged ions), and the **anode**, the positive electrode which attracts **anions** (negatively charged ions).

> **Key definitions**
> **Electrolysis** is the decomposition of an ionic compound, when molten or in aqueous solution, by the passage of an electric current.

Metals or hydrogen are formed at the cathode and non-metals (other than hydrogen) are formed at the anode.

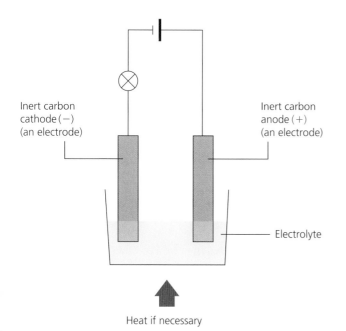

▲ **Figure 5.2** The important terms used in electrolysis

In the example of electrolysis of molten lead(II) bromide you will see that the transfer of charge during electrolysis is by:

» the movement of electrons in the metallic or graphite electrodes
» the loss or gain of electrons from the external circuit at the electrodes
» the movement of ions in the electrolyte.
The conduction that takes place in the electrodes is due to the movement of delocalised electrons (p. 46), whereas in the electrolyte the charge carriers are ions.

Electrolysis is very important within many manufacturing industries. To help you understand how aluminium sheet was obtained from its ore, we will first consider the electrolysis of lead(II) bromide.

5.2 Electrolysis of lead(II) bromide (teacher demonstration)

This experiment should only ever be carried out as a teacher demonstration and in a fume cupboard.

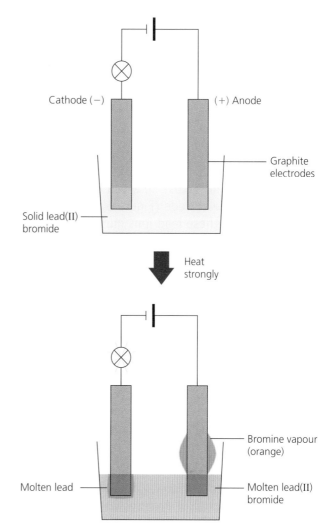

▲ **Figure 5.3** The electrolysis of molten lead(II) bromide

Consider Figure 5.3, which shows solid lead(II) bromide ($PbBr_2$) in a crucible with two carbon electrodes in contact with it. When the electrodes are first connected, the bulb does not light, because the solid compound does not allow electricity to pass through it. However, when the compound is heated until it is molten, the bulb does light. The lead(II) bromide is now behaving as an electrolyte. When this happens an orange-red vapour is seen at the anode and lead metal is produced at the cathode. Lead(II) bromide is a **binary compound**. Binary compounds are those that contain two elements chemically combined. Generally, when molten binary compounds such as lead(II) bromide are electrolysed, metals, such as lead, are formed at the cathode and non-metals, such as bromine, are formed at the anode.

The passage of an electric current in the process of electrolysis results in the break-up (decomposition) of lead(II) bromide into its constituent elements.

molten lead(II) bromide → bromine + lead

$$PbBr_2(l) \rightarrow Br_2(g) + Pb(l)$$

For lead metal to be formed, or deposited, at the cathode, the lead ions must be attracted to and move towards the electrode (Figure 5.4). To produce lead metal atoms, these lead ions must each collect two electrons at the cathode:

lead ion + electrons → lead atom

$$Pb^{2+}(l) + 2e^- \rightarrow Pb(l)$$

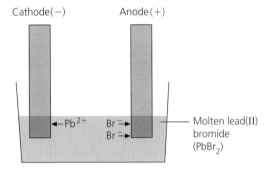

▲ **Figure 5.4** The lead ions (Pb^{2+}) are attracted to the cathode and the bromide ions (Br^-) are attracted to the anode

As you saw in Chapter 3 (p. 31), this process of gaining electrons is called reduction. Chemical equations such as those shown for the production of lead at the cathode are known as ionic half-equations. Similar ones can be written for the process taking place at the anode.

To form bromine molecules each bromide ion must first move to the anode and lose its extra negative charge at the anode, and so form a neutral bromine atom:

bromide ion → bromine atom + electron

$$Br^-(l) \rightarrow Br + e^-$$

Two bromine atoms then combine to form a bromine molecule:

bromine atoms → bromine molecule

$$2Br \rightarrow Br_2(g)$$

As you saw in Chapter 3 (p. 31), this process of losing electrons is called oxidation.

1 a Predict the products of the electrolysis of molten magnesium fluoride (MgF_2).
The metal element of the magnesium fluoride is produced at the cathode. The magnesium ions (Mg^{2+}) are attracted and deposited at this electrode.
The non-metal element of the magnesium fluoride is produced at the anode. The fluoride ions (F^-) are attracted and deposited at this electrode.

   b Write ionic half-equations for the substances produced at the different electrodes you answered in **1a**.
Ions present in MgF_2 are Mg^{2+} and F^-.
At the cathode:

$$Mg^{2+}(l) + 2e^- \rightarrow Mg(l)$$

At the anode:

$$2F^-(l) \rightarrow F_2(g) + 2e^-$$

   c State whether the processes are oxidation or reduction.
The cathode reaction is reduction (Mg^{2+} gains electrons).
The anode reaction is oxidation (F^- loses electrons).

> **Test yourself**

1 Copy and complete the following sentence about electrolysis using words from this list:

   compound   electricity   inert   chemical
   platinum   molten   breakdown

   Electrolysis is the _____ of an ionic _____ when _____ or in aqueous solution by the passage of _____.
   Electrodes are often made of graphite or _____ and are generally used in electrolysis because they are _____.

2 a Predict the products of the electrolysis of molten:
      i   potassium chloride, KCl
      ii lead(II) oxide, PbO.

   b Write ionic half-equations for the substances produced at the different electrodes for **a i** and **a ii**.

   c Use the answers to **2b** to decide which processes are oxidation and which are reduction.

5.3 Electrolysis of aluminium oxide

Aluminium is the most abundant metallic element in the Earth's crust. The main **ore** of aluminium is bauxite (Figure 5.5) and aluminium is extracted from it by electrolysis. Aluminium is a reactive metal so it is very difficult to extract from its ore. Reactive metals hold on tightly to the element(s) they have combined with and many are extracted from their ores by electrolysis.

▲ **Figure 5.5** Bauxite mining

Today we use aluminium in very large quantities. The commercial extraction of aluminium involves the following stages.

›› Bauxite, an impure form of aluminium oxide, is first treated to obtain pure aluminium oxide. This improves the conductivity of the molten aluminium oxide.

›› The purified aluminium oxide is then dissolved in molten cryolite (Na_3AlF_6). Cryolite, a mineral found naturally in Greenland, is used to reduce the working temperature of the electrolysis cell from 2017°C (the melting point of pure aluminium oxide) to between 800 and 1000°C. Therefore, the cryolite provides a very large saving in the energy requirements of the process.

In recent years it has become necessary to manufacture the cryolite.

» The molten mixture is then electrolysed in a cell similar to that shown in Figure 5.6.

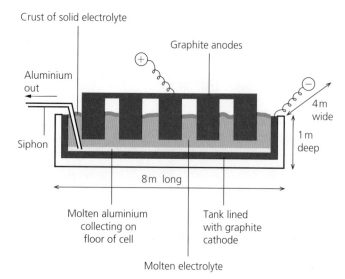

▲ **Figure 5.6** The cell is used in industry to extract aluminium by electrolysis

The anodes of this process are blocks of carbon (graphite) which are lowered into the molten mixture from above. The cathode is the graphite lining of the steel vessel containing the cell.

Aluminium oxide is an ionic compound. When it is melted the ions become mobile, as the strong electrostatic forces of attraction between them are broken by the input of heat energy. During electrolysis the negatively charged oxide ions are attracted to the anode (the positive electrode), where they lose electrons (oxidation).

oxide ions → oxygen molecules + electrons

$$2O^{2-}(l) \rightarrow O_2(g) + 4e^-$$

The positive aluminium ions are attracted to the cathode (the negative electrode). They gain electrons to form molten aluminium metal (reduction).

aluminium ions → electrons + aluminium metal

$$Al^{3+}(l) \rightarrow 3e^- + Al(l)$$

A handy way of remembering what happens is **OIL RIG** – **O**xidation **I**s **L**oss, **R**eduction **I**s **G**ain of electrons.

The overall reaction which takes place in the cell is:

aluminium oxide $\xrightarrow{\text{electrolysis}}$ aluminium + oxygen

$$2Al_2O_3(l) \rightarrow 4Al(l) + 3O_2(g)$$

The molten aluminium collects at the bottom of the cell and it is siphoned out at regular intervals (Figure 5.6). No other metals are deposited, since the cryolite is mostly not affected by the flow of electricity. Problems do arise, however, with the graphite anodes. At the working temperature of the cell, the oxygen liberated reacts with the graphite anodes, producing carbon dioxide.

carbon + oxygen → carbon dioxide

$$C(s) + O_2(g) \rightarrow CO_2(g)$$

The anodes burn away and have to be replaced on a regular basis.

The electrolysis of aluminium oxide is a continuous process in which vast amounts of electricity are used. In order to make the process an economic one, a cheap form of electricity is required. Hydroelectric power (HEP) is usually used for this process. The plant shown in Figure 5.7 uses an HEP scheme to provide some of the electrical energy required for this process.

▲ **Figure 5.7** An aluminium smelting plant

Using cheap electrical energy has allowed aluminium to be produced in such large quantities that it is the second most widely used metal after iron. It is used in the manufacture of electrical/transmission cables. Aluminium strands are wrapped around a core of steel wires. Aluminium is used because of its low density, chemical inertness and good **electrical conductivity**. This means that it will take a high voltage and low current to carry the same power. Owing to the first two of these properties of aluminium, it is also used for making cars, bikes, cooking foil and food containers, as well as in alloys (Chapter 10, p. 159) such as duralumin, which is used in the manufacture of aeroplane bodies (Figure 5.8). Worldwide production of aluminium now exceeds 40 million tonnes each year.

▲ **Figure 5.8** Aluminium is used in the manufacture of aeroplane bodies

---

⮕ Going further

Recycling aluminium

The metal aluminium is the third most abundant element in the Earth's crust after the non-metals oxygen and silicon, and the most abundant metal in the Earth's crust. However, it is more expensive to produce than many metals, such as iron, due to the cost of the large amounts of electricity which are needed for the electrolysis process. As a result, globally up to 50% of the aluminium used is recycled.

Obtaining aluminium from recycled scrap requires approximately 5% of the energy needed to extract aluminium from its ore. However, collecting scrap metal and transport to the metal extraction plant adds extra costs to the overall recycling costs.

There are serious environmental problems associated with the location of aluminium plants including:

» the effects of the extracted impurities, which form a red mud (Figure 5.9)
» the fine cryolite dust, which is emitted through very tall chimneys so as not to affect the surrounding area
» the claimed link between environmental aluminium and a degenerative brain disorder called Alzheimer's disease – it is thought that aluminium is a major influence on the early onset of this disease. However, the evidence is still inconclusive.

▲ **Figure 5.9** Pollution from bauxite mining in Malaysia

▶ Test yourself

3 Produce a flow chart to summarise the processes involved in the extraction of aluminium metal from bauxite.
4 Explain why the mixture of gases formed at the anode contains oxygen, carbon dioxide and traces of fluorine.

Anodising

Anodising is an electrolytic process in which the surface coating of oxide on aluminium (Al_2O_3) is made thicker. In this process the aluminium object is made the anode in a cell in which the electrolyte is dilute sulfuric acid. During the electrolysis process, oxygen is produced at the anode and combines with the aluminium. The oxide layer on the surface of the aluminium therefore increases. Dyes can be mixed with the electrolyte and so the new thicker coating of oxide is colourful and also decorative (Figure 5.10).

▲ **Figure 5.10** The oxide layer on the surface of these aluminium cups has been thickened, and dyes added to obtain the vibrant colours

Test yourself

5 Complete the following passage about the extraction of aluminium using words from the list below.

anode   cell   cryolite   replaced   burn
cathode   oxygen   bauxite   carbon

Aluminium is extracted from its molten ore, _____ by electrolysis in an electrolytic _____. Before the extraction process begins _____ is added. The _____ electrodes are then lowered into the cell. During the process aluminium is produced at the _____ and is siphoned off. At the _____ oxygen is produced. Because of the high temperature required in the process the carbon electrodes _____ away due to the reaction of the anodes with _____ produced at that electrode. Therefore the anodes have to be _____ regularly.

5.4 Electrolysis of aqueous solutions

Other industrial processes involve the electrolysis of aqueous solutions. To explain what is happening in these processes, we will first consider the electrolysis of dilute sulfuric acid.

Electrolysis of dilute sulfuric acid

Pure water is a very poor conductor of electricity because there are so few ions in it. However, it can be made to decompose if an electric current is passed through it in a Hofmann voltameter, as in Figure 5.11.

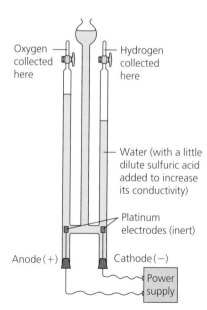

Oxygen collected here

Hydrogen collected here

Water (with a little dilute sulfuric acid added to increase its conductivity)

Platinum electrodes (inert)

Anode (+)   Cathode (−)

Power supply

▲ **Figure 5.11** A Hofmann voltameter used to electrolyse water

To enable water to conduct electricity better, some dilute sulfuric acid (or sodium hydroxide solution) is added. When the power is turned on and an electric current flows through this solution, gases can be seen to be produced at the two electrodes and they are collected in the side arms of the apparatus. After about 20 minutes, approximately twice as much gas is produced at the cathode as at the anode.

The gas collected at the cathode burns with a squeaky pop, showing it to be hydrogen gas (Table 14.6, p. 227).

For hydrogen to be collected in this way, the positively charged hydrogen ions must have moved to the cathode.

hydrogen ions + electrons → hydrogen molecules

$$4H^+(aq) \quad + \quad 4e^- \quad \rightarrow \quad 2H_2(g)$$

If during this process the water molecules lose $H^+(aq)$, then the remaining portion must be hydroxide ions, $OH^-(aq)$. These ions are attracted to the anode.

The gas collected at the anode relights a glowing splint, showing it to be oxygen (Table 14.6, p. 227).

This gas is produced in the following way.

hydroxide ions → water + oxygen + electrons

$$4OH^-(aq) \quad \rightarrow 2H_2O(l) + \ O_2(g) + \quad 4e^-$$

Generally when aqueous solutions are electrolysed, hydrogen is produced at the cathode and non-metals (other than hydrogen) are formed at the anode.

This experiment was first carried out by Sir Humphry Davy. It confirmed that the formula for water was H_2O.

In the electrolysis of dilute sulfuric acid, platinum (an inert electrode) may be replaced by carbon (graphite). The only difference to occur is that, as well as oxygen being produced at the anode, a little carbon dioxide will also be formed.

Electrolysis of concentrated aqueous sodium chloride

When sodium chloride (NaCl) is dissolved in water the ions present become free to move. When concentrated aqueous sodium chloride is electrolysed the two gases, hydrogen and chlorine, are produced. Sodium might have been expected to be produced at the cathode. Why does this not happen?

Four ions are present in the solution:

» from water: H^+ and OH^-
» from sodium chloride: Na^+ and Cl^-

When the current flows the H^+ ions and the Na^+ ions are both attracted to the cathode. The H^+ ions, however, accept electrons more easily than the Na^+ ions and so hydrogen gas (H_2) is produced at the cathode.

hydrogen ions + electrons $\xrightarrow{\text{reduction}}$ hydrogen molecules

$$2H^+(aq) \quad + \quad 2e^- \quad \rightarrow \quad H_2(g)$$

Because the sodium is more reactive than the hydrogen it loses electrons more easily. Generally it is found that during the electrolysis of aqueous solutions, very reactive metals, like sodium, are not formed at the cathode. You will see in Chapter 10 that reactive metals liberate hydrogen when they react with water.

OH^- and Cl^- are attracted to the anode. The Cl^- ions release electrons more readily than the OH^- ions and so chlorine gas is produced by the electrode process.

chloride ions $\xrightarrow{\text{oxidation}}$ chlorine molecules + electrons

$$2Cl^-(aq) \quad \rightarrow \quad Cl_2(g) \quad + \quad 2e^-$$

This leaves a high concentration of hydroxide ions, $OH^-(aq)$, around the cathode. The solution in the region of the anode therefore becomes alkaline.

Electrolysis guidelines

The following points may help you work out the products of electrolysis in unfamiliar situations. They will also help you remember what happens at each electrode.

» Non-metals are produced at the anode whereas metals and hydrogen gas are produced at the cathode.
» At an inert anode, chlorine, bromine and iodine (the halogens) are produced in preference to oxygen.
» At an inert cathode, hydrogen is produced in preference to metals unless unreactive metals such as copper are present.

Going further

The chlor-alkali industry

The electrolysis of concentrated or saturated sodium chloride solution (brine) is the basis of a major industry. In countries where rock salt (sodium chloride) is found underground, it is mined. In other countries it can be obtained by evaporation of sea water in large, shallow lakes. Three important substances are produced in this electrolysis process – chlorine, sodium hydroxide and hydrogen. The electrolytic process is very expensive, requiring vast amounts of electricity, and the process is economical only because all three products (hydrogen, chlorine and sodium hydroxide) have a large number of uses (Figure 5.12).

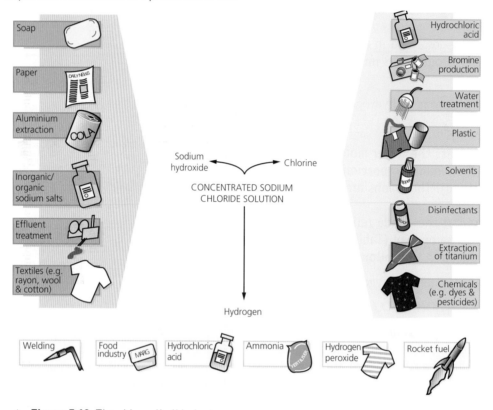

▲ **Figure 5.12** The chlor-alkali industry

? Worked example

1 Predict the products at the cathode and anode of the electrolysis of:
a dilute aqueous potassium chloride
 At cathode: hydrogen (from the water)
 At anode: chlorine and oxygen (from the water).
b concentrated hydrobromic acid (HBr)
 Concentrated acid contains very little or no water. So in the electrolysis of the acid the ions present are H⁺ and Br⁻. Hence:
 At cathode: hydrogen
 At anode: bromine.

▶ Test yourself

6 Suggest a reason why, in the electrolysis of acidified water, the volume of hydrogen gas produced at the cathode is only approximately twice the volume of oxygen gas produced at the anode.
7 The uses of sodium hydroxide can be separated on a percentage basis as follows:
 Neutralisation: 5%
 Paper manufacture: 5%
 Oil refining: 5%
 Soap/detergents: 5%
 Manufacture of rayon and acetate fibres: 16%
 Manufacture of chemicals: 30%
 Miscellaneous uses: 34%
 Use a graph-plotting program to create a chart of these data.

5.5 Electrolysis of copper(II) sulfate aqueous solution

Copper(II) sulfate solution ($CuSO_4(aq)$) may be electrolysed using inert graphite electrodes in a cell similar to that shown in Figure 5.13. When the solution is electrolysed, oxygen gas and copper metal are formed at the anode and cathode respectively. Four ions are present in solution:

» from the water: $H^+(aq) + OH^-(aq)$
» from the copper(II) sulfate: $Cu^{2+}(aq) + SO_4^{2-}(aq)$

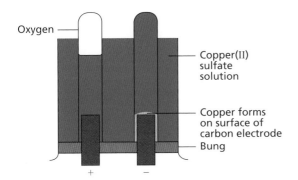

▲ **Figure 5.13** The electrolysis of copper(II) sulfate solution using inert electrodes

$H^+(aq)$ and $Cu^{2+}(aq)$ ions are both attracted to the cathode, the Cu^{2+} ions accepting electrons more readily than the H^+ ions (preferential discharge). Copper metal is therefore deposited at the cathode (Figure 5.13).

copper ions + electrons → copper atoms

$$Cu^{2+}(aq) + 2e^- \rightarrow Cu(s)$$

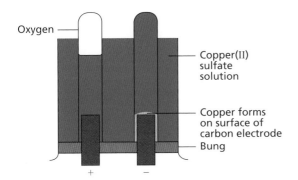

▲ **Figure 5.14** Oxygen is given off at the anode and copper is deposited at the cathode

$OH^-(aq)$ and $SO_4^{2-}(aq)$ ions are both attracted to the anode. The OH^- ions release electrons more easily

than the SO_4^{2-} ions, so oxygen gas and water are produced at the anode (Figure 5.14).

hydroxide ions → oxygen + water + electrons

$$4OH^-(aq) \rightarrow O_2(g) + 2H_2O(l) + 4e^-$$

Purification of copper

As copper is a very good conductor of electricity, it is used for electrical wiring and cables (Figure 5.15). Pure copper is also used in the manufacture of cooking utensils owing to its high thermal conductivity.

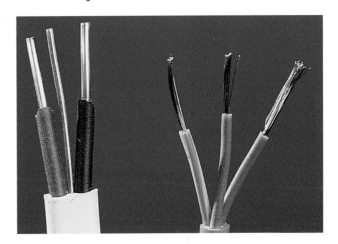

a The copper used in electrical wiring has to be very pure

b Due to the high density of copper and its cost, steel-cored aluminium cables are used for electrical transmission through national grids

▲ **Figure 5.15**

However, even small amounts of impurities cut down this conductivity quite noticeably, whether in fine wires or larger cables. The metal must be 99.99% pure to be used in this way. To ensure this level of purity, newly extracted copper has to be purified by electrolysis.

The impure copper is used as the anode in the electrolysis process, and is typically 1 m square, 35–50 mm thick and 330 kg in weight. The cathode is a 1 mm thick sheet and weighs about 5 kg; it is made from very pure copper. Because copper

is itself involved in the electrolytic process, the copper cathode is known as an 'active' electrode. The electrolyte is a solution of copper(II) sulfate (0.3 mol/dm^3) acidified with a 2 mol/dm^3 solution of sulfuric acid to help the solution conduct electricity (Figure 5.16).

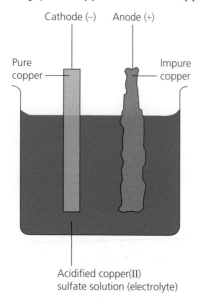

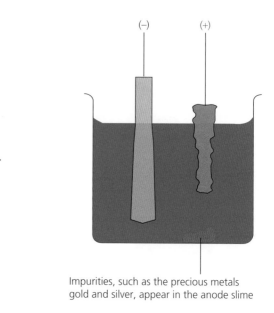

▲ **Figure 5.16** Copper purification process

When the current flows, the copper moves from the impure anode to the pure cathode. Any impurities fall to the bottom of the cell and collect below the anode in the form of a slime. This slime is rich in precious metals and the recovery of these metals is an important aspect of the economics of the process. The electrolysis proceeds for about three weeks until the anodes are reduced to about 10% of their original size and the cathodes weigh between 100 and 120 kg. A potential of 0.25 V and a current density of 200 A/m^2 are usually used.

The ions present in the solution are:

» from the water: $H^+(aq) + OH^-(aq)$
» from the copper(II) sulfate: $Cu^{2+}(aq) + SO_4^{2-}(aq)$

During the process the impure anode loses mass because the copper atoms lose electrons and become copper ions, $Cu^{2+}(aq)$ (Figure 5.17).

copper atoms → copper ions + electrons

$$Cu(s) \quad \rightarrow \quad Cu^{2+}(aq) + \quad 2e^-$$

The electrons released at the anode travel around the external circuit to the cathode. There the electrons are passed on to the copper ions, $Cu^{2+}(aq)$, from

the copper(II) sulfate solution and the copper is deposited or copper plated on to the cathode.

copper ions + electrons → copper atoms

$$Cu^{2+}(aq) + \quad 2e^- \quad \rightarrow \quad Cu(s)$$

▲ **Figure 5.17** The movement of ions in the purification of copper by electrolysis

Going further

The annual production of copper worldwide is in excess of 16 million tonnes. However, a large amount of the copper we need is obtained by recycling. This way of obtaining copper is increasing in importance as it becomes more difficult and expensive to locate and extract the copper ore from the ground.

Practical skills

Electrolysis of copper sulfate solution

Safety

- Eye protection must be worn.
- Take care with copper(II) sulfate solution: ≤ 1.0 mol/dm^3 is an irritant and corrosive (Hazard).

As part of a project about purifying metals a student wanted to investigate the purification of copper. Their teacher gave them a piece of impure copper and a piece of pure copper foil, as well as some sulfuric acid acidified copper(II) sulfate solution. The apparatus available is shown in Figure 5.16. The electrical energy was provided by a power supply. The student then followed these directions:

1. Weigh the copper pieces separately and enter the masses in your table of results. You should also record your observations after the process is complete.
2. Set up the apparatus as shown in Figure 5.16, adding 50 cm^3 of the acidified copper(II) sulfate to the beaker.
3. Switch on the power supply and let the process run for 15 minutes.
4. Switch off the power supply, carefully dry the copper strips and reweigh the copper strips.

This is the table of results obtained by the student.

Copper strip	Mass at start/g	Mass after 15 minutes/g	Change in mass/g	Observations
Pure copper cathode (–)	3.16	3.40		Has an orange/brown colouration on the part of the copper that has been in the copper(II) sulfate solution
Impure copper anode (+)	3.25	2.98		Looks a little thinner. Small pile of material below this electrode

a What safety precautions would you take in carrying out the experiment?

b Why do you think the copper(II) sulfate was acidified?

c The colour of the copper(II) sulfate solution does not change. Why not?

d Calculate the mass change for each of the copper electrodes.

e Are the changes in mass what you expected? Explain your answer.

f It was found that there was some material below the anode. Explain how you could obtain a sample of this material.

g It was thought that this mixture you have isolated in **f** was a mixture of substances including what appeared to be a dye.
   i Which technique could be used to check this out?
   ii Design an experiment, using the technique you have described in **i**, which would allow you to identify the components of the dye material.

> ### ► Test yourself
>
> 8 Why do you think it is advantageous to use inert electrodes in the electrolysis processes?
> 9 Predict the products of electrolysis of a solution of copper(II) sulfate if carbon electrodes are used instead of those made from copper as referred to in the purification of copper section.
> 10 Predict the products of the electrolysis of concentrated hydrochloric acid using platinum electrodes. (Remember concentrated acids have very little or no water present.)
> 11 Using your knowledge of electrolysis, predict the likely products of the electrolysis of copper(II) chloride solution, using platinum electrodes. Write ionic half-equations for the formation of these products at the electrodes.

5.6 Fuel cells

Scientists have found a way of changing chemical energy into electrical energy, using a **fuel cell** (Figure 5.18). Fuel cells are chemical cells except that the reagents are supplied continuously to the electrodes. The reagents are usually hydrogen and oxygen. The fuel cell principle was first discovered by Sir William Grove in 1839.

▲ **Figure 5.18** Fuel cell electric vehicles are being developed and may replace petrol and diesel engines

When Grove was electrolysing water and he switched off the power supply, he noticed that a current still flowed but in the reverse direction. Subsequently,

the process was explained in terms of the reactions at the electrodes' surfaces of the oxygen and hydrogen gases which had been produced during the electrolysis.

The hydrogen fuel cells used to power electric motors are about 70% efficient and, since the only product is water, they are pollution free.

> **Key definition**
>
> A hydrogen–oxygen **fuel cell** uses hydrogen and oxygen to produce electricity with water as the only substance produced.

The aqueous NaOH electrolyte is kept within the cell by electrodes which are porous, allowing the transfer of O_2, H_2 and water through them (Figure 5.19). As O_2 gas is passed into the cathode region of the cell it is reduced:

$$O_2(g) + 2H_2O(l) + 4e^- \rightarrow 4OH^-(aq)$$

The OH^- ions formed are removed from the fuel cell by reaction with H_2:

$$H_2(g) + 2OH^-(aq) \rightarrow 2H_2O(l) + 2e^-$$

The electrons produced by this process pass around an external circuit to the cathode.

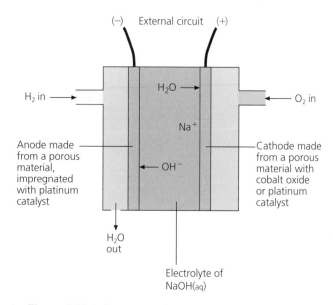

▲ **Figure 5.19** A diagrammatic view of a fuel cell

Advantages and disadvantages of the hydrogen fuel cell

As issues surrounding climate change gain momentum, people and governments all around the world are growing more and more concerned each day. Countries all over the world are considering cleaner and more sustainable sources of power. There is adoption of renewable technologies, such as electric vehicles. Another technology that is gaining interest worldwide is fuel cell technology. Fuel cell technology has its advantages. For example, a fuel cell:

» uses hydrogen and oxygen and makes non-polluting water in the process of generating electricity, whereas petrol and diesel engines produce many pollutants
» is similar to a battery but does not require any external charging
» is capable of producing electricity as long as hydrogen fuel and oxygen are supplied.

However there are some disadvantages to using fuel cells. For example:

» hydrogen is in the gas state at room temperature and pressure, so it is difficult to store in a car
» the infrastructure does not yet exist, as it does for fossil fuels, for example, the number of refuelling stations
» fuel cells and electric motors are less durable than petrol or diesel engines, which means they do not last as long
» fuel cells are very expensive at the moment.

> **Test yourself**
>
> **12** The fuel cell was discovered during electrolysis experiments with water. It is the reverse process which produces the electricity. Write a balanced chemical equation to represent the overall reaction taking place in a fuel cell.

5.7 Electroplating

Electroplating is the process involving electrolysis where one metal is plated, or coated, with another. Often the purpose of **electroplating** is to improve the appearance of the object and give a protective coating to the metal beneath. For example, bath taps are chromium plated to prevent corrosion, and at the same time are given a shiny, more attractive finish (Figure 5.20).

> **Key definition**
>
> **Electroplating** is applied to metals to improve their appearance and resistance to corrosion.

▲ **Figure 5.20** This tap has been chromium plated

The electroplating process is carried out in a cell such as the one shown in Figure 5.21a. This is often known as the 'plating bath' and it contains a suitable electrolyte, usually a solution of a metal salt.

For silver plating the electrolyte is a solution of a silver salt. The item to be plated is made the cathode in the cell so that the metal ions move to it when the current is switched on. The cathode reaction in this process is:

silver ions + electrons → silver atoms

$$Ag^+(aq) \ + \ e^- \ \rightarrow \ Ag(s)$$

Anode (+)   Cathode (−)

Object to be plated,
e.g. spoon

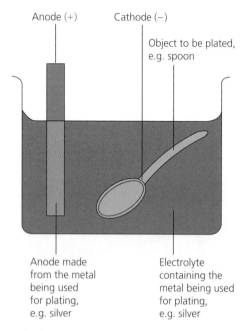

Anode made
from the metal
being used
for plating,
e.g. silver

Electrolyte
containing the
metal being used
for plating,
e.g. silver

a Silver plating a spoon

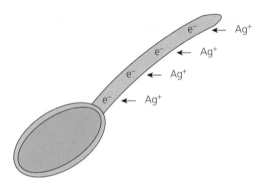

The Ag⁺ ions are attracted to the
cathode, where they gain electrons

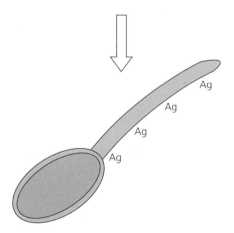

A coating of silver forms
on the spoon at the cathode

b Explaining silver plating

▲ **Figure 5.21**

 Going further

Plating plastics

Nowadays it is not only metals that are electroplated. Plastics have been developed that are able to conduct electricity. For example, the plastic poly(pyrrole) can be electroplated in the same way as the metals we have discussed earlier (Figure 5.22).

▲ **Figure 5.22** This plastic has been coated with copper by electrolysis

▲ **Figure 5.23** This leaf has been electroplated

Test yourself

13 Explain why copper(II) chloride solution would not be used as an electrolyte in the electrolyte cell used for copper plating.

14 Write equations which represent the discharge at the cathode of the following ions:
  a K^+ b Pb^{2+} c Al^{3+}
  and at the anode of:
  d Br^- e O^{2-} f F^-

Revision checklist

After studying Chapter 5 you should be able to:

✓ Define electrolysis.
✓ Identify the component parts of anode, cathode and the electrolyte in a simple electrolytic cell.
✓ Identify which electrode in a simple electrolytic cell the anions and cations are attracted to.
✓ Describe the transfer of charge that takes place during electrolysis.
✓ Construct ionic half-equations for reactions at the anode (to show oxidation) and at the cathode (to show reduction).
✓ Describe the electrode products as well as the observations made during the electrolysis of the following substances: molten lead(II) bromide, concentrated aqueous sodium chloride, dilute sulfuric acid.
✓ Describe the process of the extraction of aluminium from purified bauxite.

✓ State that metals or hydrogen are formed at the cathode and that non-metals (other than hydrogen) are formed at the anode.
✓ Predict the identity of the products of electrolysis of a specified binary compound, such as lead(II) bromide, in the molten state.
✓ Describe and identify the products of electrolysis using an aqueous solution of copper(II) sulfate with carbon and copper electrodes.
✓ Describe how and why metals are electroplated.
✓ State that hydrogen–oxygen fuel cells use hydrogen and oxygen to generate electricity with water as the only substance produced.
✓ Describe the advantages and disadvantages of using hydrogen–oxygen fuel cells in comparison with gasoline or petrol engines in vehicles.

Exam-style questions

1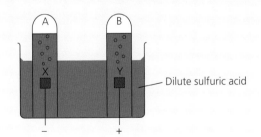

— Dilute sulfuric acid

This is a diagram of an experiment in which electricity was passed through a mixture of distilled water containing a little dilute sulfuric acid.
  a Identify the gas that collects at **A**. [1]
  b Identify the gas that collects at **B**. [1]

  c If 100 cm³ of gas collects in **A**, how much would there be in **B**? [1]
  d Identify the metal usually used for **X** and **Y**. [1]
  e **X** is called the _____ . [1]
  f **Y** is called the _____ . [1]
  g Give the formulae of the three ions present in the solution. [3]
  h Give the ionic half-equations for the reactions that take place at both X and Y. [4]

2 Explain the meaning of each of the following terms.
  a Anode [2]   e Anion [2]
  b Cathode [2]   f Cation [2]
  c Electrolysis [2]   g Oxidation [1]
  d Electrolyte [2]   h Reduction [1]

3 Copper is purified by electrolysis, as in the example shown below.

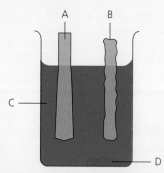

a Identify the materials used for the electrodes **A** and **B**. [2]

b Identify the electrolyte **C** and substance **D**. [2]

c Explain why substance **D** is of economic importance in respect of this process. [2]

d Give word and symbol equations for the reactions which take place at the cathode and anode during this process. [6]

e Explain why electrolyte **C** has to be acidified with dilute sulfuric acid. [2]

f Explain why copper has to be 99.99% pure for use in electrical cables. [2]

4 Copy and complete the table below, which shows the results of the electrolysis of four substances using inert electrodes. [5]

Electrolyte	Product at anode (positive electrode)	Product at cathode (negative electrode)
Molten aluminium oxide		Aluminium
Concentrated sodium chloride solution	Chlorine	
Molten lithium chloride		
Silver nitrate solution		Silver

a Give the meaning of 'inert electrodes'. [2]

b Explain why the sodium chloride solution becomes progressively more alkaline during electrolysis. [3]

c Explain why solid lithium chloride is a non-conductor of electricity, whereas molten lithium chloride and lithium chloride solution are good conductors of electricity. [3]

d During the electrolysis of molten aluminium chloride ($AlCl_3$) the carbon anodes are burned away. Explain why this happens and write balanced chemical equations for the reactions that take place. [4]

5 Sodium hydroxide is made by the electrolysis of brine (concentrated sodium chloride solution).

a Give ionic half-equations for the reactions which take place at the cathode and anode. State clearly whether a reaction is oxidation or reduction. [8]

b Give two large-scale uses of the products of this electrolytic process. [6]

c Analyse the following statement: 'This electrolytic process is a very expensive one'. [2]

6 Electroplating is an important industrial process.

a Explain what electroplating is. [2]

b Explain why certain metals are electroplated. [2]

c Give two examples of the use of electroplating. [2]

6 Chemical energetics

FOCUS POINTS

★ What are fossil fuels and how are they used?
★ How do we obtain a range of useful materials from petroleum?
★ How do we obtain energy from chemical reactions?

In this chapter you will look at the energy, or enthalpy, changes which happen when a chemical reaction occurs. We know that when we burn a fuel energy is released, and the surroundings get warmer. Many of the fuels we use are obtained from the fossil fuel petroleum. How are important fuels such as gasoline or diesel obtained from petroleum?

In this chapter we will look at why heat is produced and you will be introduced to the idea of chemical reactions which occur and cause things to get cooler. Have you met any of these reactions? What could they be used for?

By the end of this chapter you should be able to calculate the energy change which occurs when a chemical reaction happens.

6.1 Substances from petroleum

What do the modes of transport in Figure 6.1 have in common?

▲ **Figure 6.1** What type of fuel do these modes of transport use?

They all use liquids obtained from **petroleum** as fuels.

> **Key definition**
> **Petroleum** is a mixture of hydrocarbons.

Oil refining

Petroleum (also called crude oil) is a complex mixture of compounds known as **hydrocarbons** (Figure 6.2). Hydrocarbons are molecules which contain only the elements hydrogen and carbon bonded together covalently (Chapter 3, p. 37). These carbon compounds form the basis of a group called **organic compounds**. All living things are made from organic compounds based on chains of carbon atoms similar to those found in petroleum. Petroleum is not only a major source of fuel but also a raw material of enormous importance. It supplies a large and diverse chemical industry to make dozens of products that we use in everyday life, such as polymers and medicines.

> **Key definition**
>
> **Hydrocarbons** are compounds that contain hydrogen and carbon only.

▲ **Figure 6.3** Oil drilling rig off Labuan island, northwest coast of Borneo

▲ **Figure 6.2** Petroleum is a mixture of hydrocarbons

Petroleum is not very useful to us until it has been processed. The process, known as **oil refining**, is carried out at an oil refinery.

Refining involves separating petroleum into various batches or **fractions**. Chemists use a technique called **fractional distillation** to separate the different fractions. This process works in a similar way to that discussed in Chapter 14, p. 220, for separating ethanol (alcohol) and water. The different components (fractions) separate because they have different boiling points. The petroleum is heated to about 400°C to vaporise all the different parts of the mixture. The mixture of vapours passes into the

fractionating column near the bottom (Figure 6.4). Each fraction is obtained by collecting hydrocarbon molecules, which have a boiling point in a given range of temperatures. For example, the fraction we know as petrol contains molecules which have boiling points between 30°C and 110°C. The molecules in this fraction contain between five and ten carbon atoms (short hydrocarbon chains). These smaller molecules with the shorter chain lengths have lower boiling points and condense higher up the tower. Bigger hydrocarbon molecules (long hydrocarbon chains), with longer chain lengths, have higher boiling points and condense in the lower half of the tower.

The liquids condensing at different levels are collected on trays. In this way the petroleum is separated into different fractions. This process is known as fractionation. These fractions usually contain a number of different hydrocarbons. The individual single hydrocarbons can then be obtained by refining the fraction by further distillation.

It is important to realise that the uses of the fractions depend on their properties, such as **viscosity** and **volatility**. For example, one of the lower fractions, which boils in the range 250–350°C, is quite viscous (does not flow easily). Because it is thick and sticky, it makes a good lubricant. The fractions that are volatile (evaporate easily) are easy to ignite. The gasoline or petrol fraction burns very easily, which makes it a good fuel for use in engines. Figure 6.4b shows the uses of the fractions.

In general, the properties of the fractions obtained from this fractional distillation change from the bottom to the top of the tower with:

» lowering boiling points
» higher volatility
» lower viscosity
» decreasing chain length.

a Fractional distillation of petroleum in a refinery

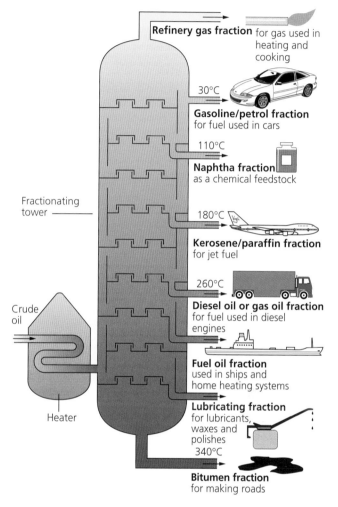

b Uses of the different fractions obtained from petroleum

▲ **Figure 6.4**

Test yourself

1 What do you understand by the term hydrocarbon?
2 All living organisms are composed of compounds which contain carbon. Why do you think carbon chemistry is often called 'organic chemistry'?
3 List the main fractions obtained by separating the petroleum mixture and explain how they are obtained in a refinery.
4 Which of the fractions obtained from petroleum could be used for producing:

   a waxes
   b fuel for aeroplanes
   c polymers?

5 Describe how the properties of fractions obtained from petroleum change as they are collected in a fractionating column from the bottom to the top of the column, in terms of their:

   a chain length   c viscosity
   b volatility   d boiling points.

6.2 Fossil fuels

Coal, petroleum and natural gas are all examples of **fossil fuels**. The term 'fossil fuels' is derived from the fact that they are formed from dead plants and animals which were fossilised over 200 million years ago (Figure 6.5).

Going further

Formation of fossil fuels

Coal was produced by the action of pressure and heat on dead wood from ancient forests. When dead trees fell into the swamps they were buried by mud. Over millions of years, due to movement of the Earth's crust and to changes in climate, the land sank and the decaying wood became covered by even more layers of mud and sand. As time passed the gradually forming coal became more and more compressed as other material was laid down above it (Figure 6.5).

Petroleum and gas were formed during the same period as coal. It is believed petroleum and gas were formed from the remains of plants, animals and bacteria that once lived in seas and lakes. This material sank to the bottom of these seas and lakes and became covered in mud, sand and silt which thickened with time.

Anaerobic decay (decay which takes place in the absence of oxygen) took place and, as the mud layers built up, high temperatures and pressures were created which converted the material slowly into petroleum and gas. After the rock formed, Earth movements caused it to bend and split, and the oil and gas were trapped in folds beneath layers of non-porous rock or cap-rock (Figure 6.6).

▲ **Figure 6.5** Piece of coal showing a fossilised leaf

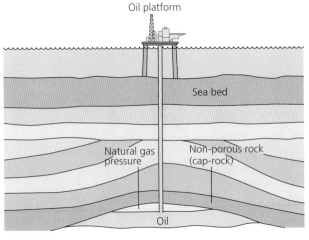

▲ **Figure 6.6** Natural gas and petroleum are trapped under non-porous rock

Test yourself

6 Coal, petroleum and natural gas are all termed 'fossil fuels'. Why is the word 'fossil' used in this context?

7 Draw a flow diagram to represent the formation of coal, oil or gas.

6.3 What is a fuel?

A fuel is any substance which can be conveniently used as a source of energy. Fossil fuels release energy in the form of heat when they undergo **combustion**.

$$\text{fossil fuel} + \text{oxygen} \rightarrow \text{carbon} + \text{water}$$
$$\text{dioxide}$$

For example, natural gas burns readily in air (Chapter 12, p. 185).

$$\text{methane} + \text{oxygen} \rightarrow \text{carbon} + \text{water}$$
$$\text{dioxide}$$

$$CH_4(g) + 2O_2(g) \rightarrow CO_2(g) + 2H_2O(l)$$

Natural gas, like petroleum, is a mixture of hydrocarbons. Methane is the main constituent of natural gas; it also contains ethane and propane, and may also contain some sulfur. The sulfur content varies from source to source: natural gas obtained from Oman and the United Arab Emirates is quite low in sulfur.

The perfect fuel would:

» be cheap
» be available in large quantities
» be safe to store and transport
» be easy to ignite and burn, causing no pollution
» release no greenhouse gases
» be capable of releasing large amounts of energy.

Solid fuels are safer than volatile liquid fuels, like petrol, and gaseous fuels, like natural gas.

Test yourself

8 'We have not yet found the perfect fuel'. Discuss this statement.

6.4 Alternative sources to fossil fuels

Fossil fuels are an example of **non-renewable** resources, so called because they are not being replaced at the same rate as they are being used up. It is important to use non-renewable fuels very carefully and to consider alternative, **renewable** sources of energy for use now and in the future. These include:

» nuclear
» hydroelectric
» biomass and biogas
» hydrogen
» wind
» tidal
» wave
» geothermal
» solar.

Test yourself

9 Choose two alternative sources of energy and suggest some advantages and disadvantages of each.
10 What is meant by the terms:
   a non-renewable energy sources?
   b renewable energy sources?

6.5 Exothermic and endothermic reactions

We obtain our energy needs from the combustion of fuels.

Combustion

When natural gas burns in a plentiful supply of air it produces a large amount of energy.

methane + oxygen → carbon + water
                    dioxide

$$CH_4(g) + 2O_2(g) \rightarrow CO_2(g) + 2H_2O(l)$$

During this process, the **complete combustion** of methane, heat is given out. It is an **exothermic** reaction. An exothermic reaction is one which transfers thermal energy to the surroundings when the reaction occurs. If only a limited supply of air is available then the reaction is not as exothermic and the toxic gas carbon monoxide is produced.

> **Key definition**
>
> An **exothermic** reaction transfers thermal energy to its surroundings, leading to an increase in the temperature of the surroundings.

methane + oxygen → carbon + water
                    monoxide

$$2CH_4(g) + 3O_2(g) \rightarrow 2CO(g) + 4H_2O(l)$$

This is the **incomplete combustion** of methane.

The energy changes that take place during a chemical reaction can be shown by a **reaction pathway diagram**. Figure 6.7 shows the reaction pathway diagram for the exothermic reaction of the complete combustion of methane. The majority of chemical reactions are exothermic reactions: they give out thermal energy when they occur.

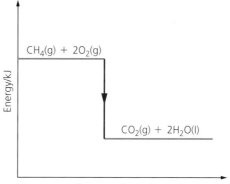

▲ **Figure 6.7** A reaction pathway diagram for the complete combustion of methane

There are also reactions which take in thermal energy and these are known as **endothermic reactions**. When an endothermic reaction occurs, energy is taken in from the surroundings and the temperature of the surroundings decreases. An example of a use of an endothermic reaction is when a cold pack is used to treat an injury. The cold pack contains an inner bag of water and a chemical which will react with it to produce an endothermic reaction. To use the cold pack you squeeze it, which allows the water to react with the chemical and, when the pack is applied to the sprain, the temperature decreases.

> **Key definition**
>
> An **endothermic** reaction transfers thermal energy from the surroundings leading to a decrease in the temperature of the surroundings.

When any reaction occurs, the chemical bonds in the reactants have to be broken – this requires energy. This is an endothermic process. When the new bonds in the products are formed, energy is given out (Figure 6.8). This is an exothermic process. The **bond energy** is defined as the amount of energy in kilojoules (kJ) associated with the breaking or making of 1 mole of chemical bonds in a molecular element or compound.

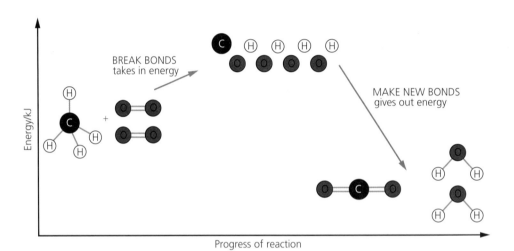

▲ **Figure 6.8** Breaking and forming bonds during the combustion of methane

Using the data in Table 6.1, which tells us how much energy is needed to break a chemical bond and how much is given out when it forms, we can calculate how much energy is involved in each stage.

▼ **Table 6.1** Bond energy data

Bond	Bond energy kJ/mol
C—H	435
O=O	497
C=O	803
H—O	464
C—C	347
C—O	358

Bond breaking

The first step is to calculate the amount of energy needed to break the bonds in the chemical reactants, methane and oxygen gas. In this reaction, 1 mole of methane reacts with 2 moles of oxygen gas.

Breaking 4 C–H bonds in 1 mole of methane requires:

$4 \times 435 = 1740 \text{ kJ}$

Breaking 2 O=O bonds in the 2 moles of oxygen requires:

$2 \times 497 = 994 \text{ kJ}$

Total = 2734 kJ of energy

This is the total amount of energy required to break the bonds in the reactants.

Making bonds

The second step is to calculate the amount of energy which will be given out when the products are made. In this reaction, 1 mole of carbon dioxide and 2 moles of water are produced.

Making two C=O bonds in 1 mole of carbon dioxide gives out:

$2 \times 803 = 1606$ kJ

Making four H–O bonds in the 2 moles of water gives out:

$4 \times 464 = 1856$ kJ

Total $= 3462$ kJ of energy

This is the total amount of energy given out when the bonds in the products are formed.

energy difference = energy required to break bonds
− energy given out when bonds
are made

$= 2734 - 3462$

$= -728$ kJ/mol of methane burned

The negative sign shows that the chemicals are losing energy to the surroundings: it is an exothermic reaction. A positive sign would indicate that the chemicals are gaining energy from the surroundings in an endothermic reaction.

The energy stored in the bonds is called the **enthalpy** and is given the symbol H. The change in energy going from reactants to products is called the **enthalpy change of reaction** and is shown as ΔH (pronounced 'delta H').

For an exothermic reaction ΔH is negative and for an endothermic reaction ΔH is positive.

> **Key definition**
>
> The transfer of thermal energy during a reaction is called the **enthalpy change, ΔH, of the reaction**.

Bond breaking is an endothermic process and bond making is an exothermic process. The enthalpy change of a reaction is the difference between the energy required to break bonds and the energy given out when bonds are made.

When fuels, such as methane, are burned they require energy to start the chemical reaction. This minimum amount of energy is known as the **activation energy**, E_a (Figure 6.9). It is the minimum amount of energy which is needed to allow the colliding particles in the reaction mixture to form products.

> **Key definition**
>
> The **activation energy**, E_a, is the minimum energy that colliding particles must have in order to react.

In the case of methane reacting with oxygen, the activation energy is some of the energy involved in the initial bond breaking, as the reaction pathway in Figure 6.9 shows. The value of the activation energy varies from fuel to fuel.

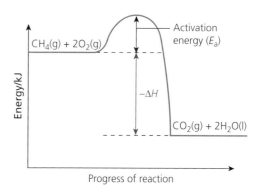

▲ **Figure 6.9** A reaction pathway diagram for methane with oxygen

Endothermic reactions are much less common than exothermic ones. In this type of reaction, energy is absorbed from the surroundings so that the energy of the products is greater than that of the reactants. The reaction between nitrogen and oxygen gases is endothermic (Figure 6.10), and the reaction is favoured (more likely to occur) at high temperatures.

nitrogen + oxygen → nitrogen(II) oxide

$N_2(g) \; + \; O_2(g) \; \rightarrow 2NO(g) \; \Delta H = +181$ kJ/mol

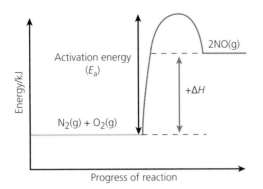

▲ **Figure 6.10** A reaction pathway diagram for nitrogen with oxygen

Dissolving is often an endothermic process. For example, when ammonium nitrate dissolves in water, the temperature of the water falls, indicating that energy is being taken from the surroundings. Photosynthesis and thermal decomposition are other examples of endothermic processes.

In equations it is usual to express the ΔH value in units of kJ/mol. For example:

$$CH_4(g) + 2O_2(g) \rightarrow CO_2(g) + 2H_2O(l); \Delta H = -728 \text{ kJ/mol}$$

This ΔH value tells us that when 1 mole of methane is burned in oxygen, 728 kJ of energy are released.

? Worked example

Look at the reaction pathway diagrams below.

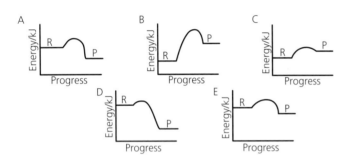

Which of the diagram(s) shows:

a an exothermic reaction

The exothermic reactions will show the products at a lower energy than the reactants, showing that energy has been given out when the reaction occurs. So the exothermic reactions are A, D and E.

b an endothermic reaction

The endothermic reactions will have the products at a higher energy than the reactants, showing that energy is being taken in, or added, when the reaction occurs. So the endothermic reactions are B and C.

c the reaction with the highest activation energy, E_a

To find which reaction has the highest activation energy draw a line across from the reactant line and then up to the top of the curve. The reaction which has the highest activation energy is the one with the longest vertical line to the top of the curve. In this example that will be reaction B.

d the most endothermic reaction

The two endothermic reactions were reactions B and C. The most endothermic reaction will have the biggest gap between the energy of the reactants and products. In this example that is reaction B.

? Worked example

Look at the two reactions below, which have been given with their ΔH values.

$$C + H_2O \rightarrow H_2 + CO; \Delta H = +131 \text{ kJ/mol}$$
$$C + O_2 \rightarrow CO_2; \Delta H = -393 \text{ kJ/mol}$$

For each of these reactions draw and label a reaction pathway diagram and label on each diagram the reactants, the products, the enthalpy change of reaction, ΔH, and the activation energy, E_a.

The first reaction is an endothermic reaction, so we need to draw a reaction pathway diagram which has the products at a higher energy than the reactants, showing that energy is being gained (or added) when the reaction occurs.

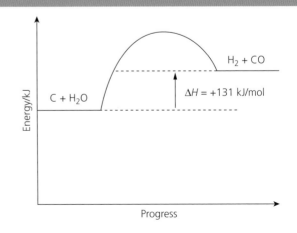

You should be able to label the ΔH value and place the symbols for the reactants and products on their lines. No information was given about the activation energy in this question so the shape and height of the curve is not specific.

The second reaction is an exothermic reaction, so we need to draw a reaction pathway diagram which has the products at a lower energy than the reactants, showing energy is being given out (lost) when the reaction occurs.

You should be able to label the ΔH value and place the symbols for the reactants and products on their lines. No information was given about the activation energy in this question so the shape and height of the curve is not specific.

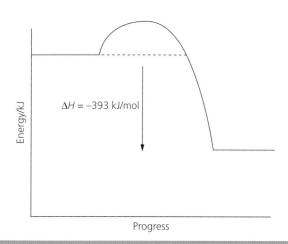

? Worked example

Using the bond energies in Table 6.1, calculate the enthalpy of reaction, ΔH, when the hydrocarbon ethane, C_2H_6, is completely combusted.

$$H-\overset{\overset{\displaystyle H}{|}}{\underset{\underset{\displaystyle H}{|}}{C}}-\overset{\overset{\displaystyle H}{|}}{\underset{\underset{\displaystyle H}{|}}{C}}-H + 3\tfrac{1}{2}O{=}O \longrightarrow 2\,O{=}C{=}O + 3\,H{\overset{O}{\diagdown}}H$$

The first step is to work out the total amount of energy needed to break the bonds in the reactants. The reactants are 1 mole of ethane and 3½ moles of oxygen gas.

Bond breaking:

Breaking 6 C–H bonds requires 6 × 435 = 2610 kJ

Breaking one C–C bond requires 347 kJ

Breaking 3½ moles of O=O bonds requires 3½ × 497 = 1739.5 kJ

Total = 4696.5 kJ

The second step is to work out the total amount of energy which is released when the bonds in the products are formed. The products are 2 moles of carbon dioxide and 3 moles of water.

Bond making:

Making 4 C=O bonds releases 4 × 803 = 3212 kJ

Making 6 H–O bonds releases 6 × 464 = 2784 kJ

Total = 5996 kJ

Enthalpy of reaction = energy required to break the bonds – energy given off when bonds are made

= 4696.5 – 5996

= –1299.5 kJ

→ Going further

Enthalpy change of neutralisation

This is the enthalpy change that takes place when 1 mole of hydrogen ions (H⁺(aq)) is neutralised.

$$H^+(aq) + OH^-(aq) \rightarrow H_2O(l); \Delta H = -57 \text{ kJ/mol}$$

This process occurs in the titration of an alkali by an acid to produce a neutral solution (Chapter 8, p. 116).

Practical skills

Experiment to find the enthalpy change of reaction

Safety

- Eye protection must be worn.
- Take care with copper(II) sulfate solution: ≤ 1.0 mol/dm^3 is an irritant and corrosive (Hazard). Disposable gloves should be worn when measuring out.

A student uses a simple experiment to find the energy given off when a chemical reaction occurs (the enthalpy change of reaction, ΔH, in the reaction between zinc metal and copper(II) sulfate. When zinc reacts with copper(II) sulfate the following reaction occurs:

zinc + copper(II) sulfate $\rightarrow$ zinc sulfate + copper

$Zn(s) + CuSO_4(aq) \rightarrow ZnSO_4(aq) + Cu(s)$

They place an expanded polystyrene cup into a beaker and pour into the expanded polystyrene cup 25 cm^3 of 1 mol/dm^3 copper sulfate solution and record the temperature of the copper(II) sulfate solution using a thermometer. They then add 5 g of zinc powder, which is in excess. The student stirred the mixture with the thermometer and recorded the highest temperature reached.

Find the change in temperature and then use the following relationship to find the thermal energy transferred during the reaction:

thermal energy transferred during the reaction

= mass of × 4.2 J/g/°C × change in
  solution   temperature/°C

1. Why is an expanded polystyrene cup used for the reaction to occur in?
2. What is the purpose of placing the expanded polystyrene cup into the beaker?
3. Why is the mixture stirred after the zinc is added?
4. Use these questions to identify improvements to this experiment.
   a. Think of possible sources of errors in this experiment.
   b. Give some improvements which will reduce these errors.

Sample data:

Initial temperature of the copper(II) sulfate solution = 19.5 °C

Maximum temperature reached = 26.9 °C

5. Use the relationship above to determine the thermal energy transferred in joules. (Assume that the density of the copper(II) sulfate solution is 1 g/cm^3.)
6. Calculate the number of moles of copper(II) sulfate used in the reaction.

You now know the amount of energy released when the number of moles you have calculated in question **6** have reacted.

7. Calculate the energy that would have been released if 1 mole of copper(II) sulfate had been used.
8. Convert the energy released into kJ and write down the value of the enthalpy change of reaction, ΔH, you have calculated.

Test yourself

11 Use the bond energy data given in Table 6.1 to answer the following questions.
   a Calculate the enthalpy change per mole for the combustion of ethanol, a fuel added to petrol in some countries.
   b Draw a reaction pathway diagram to represent this combustion process.
   c How does this compare with the enthalpy of combustion of heptane (C_7H_{16}), a major component of petrol, of -4853 kJ/mol?
   d How much energy is released per gram of ethanol and heptane burned?
12 How much energy is released if:
   a 0.5 mole of methane is burned?
   b 5 moles of methane are burned?
   c 4 g of methane are burned?
   (A_r: C = 12; H = 1)
13 How much energy is released if:
   a 2 moles of hydrogen ions are neutralised?
   b 0.25 moles of hydrogen ions are neutralised?
   c 1 mole of sulfuric acid is completely neutralised?

Revision checklist

After studying Chapter 6 you should be able to:
✓ Identify coal, natural gas and petroleum as fossil fuels, and that natural gas contains mainly methane.
✓ State that hydrocarbons are compounds that contain only the elements hydrogen and carbon.
✓ State that petroleum is a mixture of different hydrocarbon molecules.
✓ Describe the fractional distillation of petroleum to produce different fractions.
✓ Describe the changing properties of the fractions as they are produced from the fractionating column.
✓ Name uses for the fractions produced.
✓ Explain what exothermic and endothermic reactions are.
✓ Explain why energy is given out or taken in during reactions in terms of the bond breaking and bond making which occurs when the reaction happens.
✓ Interpret and draw reaction pathway diagrams showing the enthalpy change of reaction for exothermic and endothermic reactions.
✓ Define activation energy and identify this on reaction pathway diagrams.
✓ Calculate the enthalpy change of a reaction by using bond energy data.
✓ Identify coal, natural gas and petroleum as fossil fuels, and that natural gas contains mainly methane.
✓ State that hydrocarbons are compounds that contain only the elements hydrogen and carbon.
✓ State that petroleum is a mixture of different hydrocarbon molecules.

Exam-style questions

1 **a** State which of the following processes is endothermic and which is exothermic.
  i The breaking of a chemical bond. [1]
  ii The forming of a chemical bond. [1]
 b The table below shows the bond energy data for a series of covalent bonds.

Bond	Bond energy kJ/ mol
C—H	435
O=O	497
C=O	803
H—O	464
C—C	347
C—O	358

  i Use the information given in the table to calculate the overall enthalpy change for the combustion of ethanol producing carbon dioxide and water. [3]
  ii Is the process in **b i** endothermic or exothermic? [1]

2 Petroleum is a mixture of hydrocarbons. The refining of petroleum produces fractions which are more useful to us than petroleum itself. Each fraction is composed of hydrocarbons which have boiling points within a specific range of temperature. The separation is carried out in a fractionating column, as shown below.

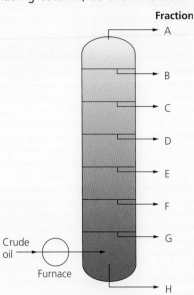

 a Which separation technique is used to separate the fractions? [1]

 b Name each of the fractions **A** to **H** and give a use for each. [8]
 c Why do the fractions come from the fractionating column in this order? [3]
 d What is the connection between your answer to **c** and the size of the molecules in each fraction? [1]
 e Which of the fractions will be the most flammable? [1]

3 Propagas is used in some heating systems where natural gas is not available. It burns according to the following equation:

$$C_3H_8(g) + 5O_2(g) \rightarrow 3CO_2(g) + 4H_2O(l);$$
$$\Delta H = -2220 \text{ kJ/mol}$$

 a What are the chemical names for propagas and natural gas?
 b Would you expect the thermal energy released per mole of propagas burned to be greater than that for natural gas? Explain your answer.
 c What is propagas obtained from?
 d Calculate:
  i the mass of propagas required to produce 5550 kJ of energy.
  ii the thermal energy produced by burning 0.5 mole of propagas.
  iii the thermal energy produced by burning 11 g of propagas.
  iv the thermal energy produced by burning 2000 dm³ of propagas.
   (A_r: H = 1; C = 12; O = 16. One mole of any gas occupies 24 dm³ at room temperature and pressure.)

4 This question is about endothermic and exothermic reactions.
 a Explain the meaning of the terms endothermic and exothermic. [2]
 b **i** Draw a reaction pathway diagram for the reaction: [4]

$$NaOH(aq) + HCl(aq) \rightarrow NaCl(aq) + H_2O(l);$$
$$\Delta H = -57 \text{ kJ/mol}$$

  ii Is this reaction endothermic or exothermic? [1]
  iii Calculate the enthalpy change associated with this reaction if 2 moles of sodium hydroxide were neutralised by excess hydrochloric acid. [1]

c i Draw a reaction pathway diagram for the reaction: [4]

$$2H_2O(l) \rightarrow 2H_2(g) + O_2(g); \Delta H = +575 \text{ kJ/mol}$$

ii Is this reaction endothermic or exothermic? [1]

iii Calculate the enthalpy change for this reaction if only 9 g of water were converted into hydrogen and oxygen. [2]

5 The following results were obtained from an experiment carried out to measure the enthalpy change of combustion of ethanol. The experiment involved heating a known volume of water with the flame from an ethanol burner.

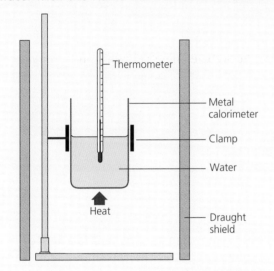

Thermometer

Metal calorimeter

Clamp

Water

Heat

Draught shield

The burner was weighed initially and after the desired temperature rise had been obtained.

Volume of water in glass beaker = 200 cm³

Mass of ethanol burner at start = 85.3 g

Mass of ethanol burner at end = 84.8 g

Temperature rise of water = 12 °C

(Density of water = 1 g/cm³)

Thermal energy transferred to water = mass of water/g × 4.2 J/g/°C × temperature rise/°C

a Calculate the mass of ethanol burned. [1]

b Calculate the amount of energy produced, in joules, in this experiment by the ethanol burning. [2]

c Convert your answer to **b** into kilojoules. [1]

d Calculate the amount of energy produced by 1 g of ethanol burning. [1]

e What is the molar mass of ethanol (C_2H_5OH)? [1]
(A_r: H = 1; C = 12; O = 16)

f How much energy would be produced if 1 mole of ethanol had been burned? (This is the enthalpy change of combustion of ethanol.) [1]

g Compare your value with the actual value of 1371 kJ/mol and suggest two reasons for the difference in values. [3]

h Write a balanced chemical equation to represent the combustion of ethanol. [1]

6 The following results were obtained from a neutralisation reaction between 1 mol/dm³ hydrochloric acid and 1 mol/dm³ sodium hydroxide. This experiment was carried out to measure the enthalpy change of neutralisation of hydrochloric acid. The temperature rise which occurred during the reaction was recorded.

Volume of sodium hydroxide used = 50 cm³

Volume of acid used = 50 cm³

Temperature rise = 5 °C

(Density of water = 1 g/cm³)

Thermal energy transferred to water = mass of water/g × 4.2 J/g/°C × temperature rise/°C

a Write a balanced chemical equation for the reaction. [1]

b What mass of solution was warmed during the reaction? [1]

c How much thermal energy was produced during the reaction? [1]

d How many moles of hydrochloric acid were involved in the reaction? [2]

e How much thermal energy would be produced if 1 mole of hydrochloric acid had reacted? (This is the enthalpy change of neutralisation of hydrochloric acid.) [2]

f The enthalpy change of neutralisation of hydrochloric acid is −57 kJ/mol. Suggest two reasons why there is a difference between this and your calculated value. [2]

Chemical reactions

FOCUS POINTS
★ How does the rate of a reaction vary?
★ Why is the rate of a chemical reaction important?
★ What is a reversible reaction?

In this chapter we will look at how we are able to alter the rate at which chemical reactions occur. Industry needs to produce chemical products in the shortest possible time, so they are made efficiently and increase profits. We know that all chemical reactions go faster when heated to a higher temperature, for example an egg cooks faster at higher temperatures, but why does this happen?

Industry also wants to make more product from a reaction. In any chemical reaction, reactants will make products eventually, so how can there be more product? We will introduce the fact that not all reactions simply produce products from reactants: some reactions are reversible and once the product has been formed, it may break down to produce reactants again.

By the end of this chapter you should be able to fully explain how the rates of chemical reactions can be altered and how reversible reactions can be manipulated to produce products efficiently.

7.1 Reactions

A physical change does not result in the formation of a new substance. It simply alters the state of the substance involved. For example, when ice melts to form water it is a physical change because the water has not changed: only its state has changed from a solid to a liquid. This is also the case if we boil water to produce steam.

A chemical change, however, results in the formation of something new. An everyday example of this is frying an egg. The chemicals in the fried egg have undergone chemical reactions forming new substances, and the fried egg cannot be changed back into the egg you started with. This is the main difference between a physical and chemical change.

In a physical change, it is possible to reverse the change, and the change can be temporary. For example, the water from melted ice can be frozen again to form ice. In a chemical change this cannot happen. For all chemical reactions, a chemical change occurs and this is a permanent change.

Figure 7.1 shows some slow and fast reactions. The two photographs at the top show examples of slow reactions. The ripening of apples takes place over a number of weeks, and the making and maturing of cheese may take months. The burning of solid fuels, such as coal, can be said to involve chemical reactions taking place at a medium speed or rate. A fast reaction is an explosion (bottom right), where the chemicals inside explosives, such as TNT, react very rapidly in reactions which are over in seconds or fractions of seconds.

As new techniques have been developed, the processes used within the chemical industry have become more complex. Therefore, chemists and chemical engineers have increasingly looked for ways to control the rates at which chemical reactions take place. In doing so, they have discovered that there are five main ways in which you can alter the rate of a chemical reaction. These ideas are not only incredibly useful to industry but can also be applied to reactions in the school laboratory.

▲ **Figure 7.1** Some slow (ripening fruit and cheese making), medium (coal fire) and fast (explosion) reactions

7.2 Factors that affect the rate of a reaction

The five main ways to alter the rate of a chemical reaction are:

» changing the concentration of solutions
» changing the pressure of gases
» changing the surface area of solids
» changing the temperature
» adding or removing a catalyst including enzymes.

Collision theory

For a chemical reaction to occur, reactant particles need to collide with one another. For products to be formed, the collision has to have a certain minimum amount of energy associated with it, and not every collision results in the formation of products. This minimum amount of energy is known as the **activation energy**, E_a (Figure 7.2). Collisions which result in the formation of products are known as successful collisions.

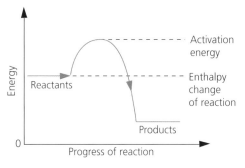

▲ **Figure 7.2** Reaction profile diagram showing activation energy

Surface area

Limestone (calcium carbonate, $CaCO_3$) is a substance which can be used to neutralise soil acidity. Powdered limestone is used as it neutralises the acidity faster than if lumps of limestone are used. Why do you think this is the case?

In the laboratory, the reaction between acid and limestone in the form of lumps or powder can be observed in a simple test-tube experiment. Figure 7.3 shows the reaction between dilute hydrochloric acid and limestone in lump and powdered form.

hydrochloric + calcium → calcium + carbon + water
acid   carbonate   chloride dioxide

$2HCl(aq) + CaCO_3(s) → CaCl_2(aq) + CO_2(g) + H_2O(l)$

▲ **Figure 7.3** The powdered limestone (left) reacts faster with the acid than the limestone in the form of lumps

The rates at which the two reactions occur can be found by measuring either:

» the volume of the carbon dioxide gas which is produced
» the loss in mass of the reaction mixture with time.

These two methods are generally used for measuring the rate of reaction for processes involving the formation of a gas as one of the products.

The apparatus shown in Figure 7.4 is used to measure the loss in mass of the reaction mixture. The mass of the conical flask plus the reaction mixture is measured at regular intervals. The total loss in mass is calculated for each reading of the balance, and this is plotted against time. Some sample results from experiments of this kind have been plotted in Figure 7.5.

The reaction between hydrochloric acid and limestone is generally at its fastest in the first minute. This is indicated by the slopes of the curves during this time. The steeper the slope, the faster the rate of reaction. You can see from the two curves in Figure 7.5 that the rate of reaction is greater with the powdered limestone than the lump form.

▲ **Figure 7.4** After 60 seconds the mass has fallen by 1.24 g

If the surface area of a reactant is increased, more particles are exposed to the other reactant, so the rate of a chemical reaction can be raised by increasing the surface area of a solid reactant.

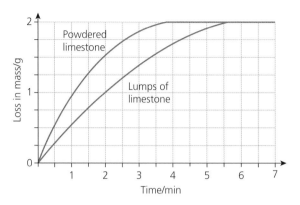

▲ **Figure 7.5** Sample results for the limestone/acid experiment

The surface area has been increased by powdering the limestone (Figure 7.6). The acid particles now have an increased amount of surface of limestone with which to collide. The products of a reaction are formed when collisions occur between reactant particles.

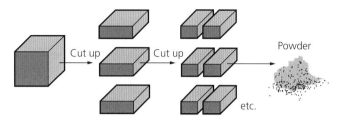

▲ **Figure 7.6** A powder has a larger surface area

Practical skills

Experiment to show the effect of surface area on reaction rate

Safety

● Eye protection must be worn.

A student took a lump of limestone (calcium carbonate) and reacted it with an excess of dilute hydrochloric acid in a conical flask which had a delivery tube attached. The gas produced was collected in an inverted burette filled with water. The experiment was repeated using smaller fragments of limestone of the same mass. The table on the right gives the results of the experiments.

1 Draw the apparatus which could be used in the experiment above.
2 Plot a graph of volume of carbon dioxide collected (*y*-axis) against time (*x*-axis) for both experiments. Use the same axes.
3 a Which of the experiments is the fastest?
    b How can you tell this from the graph?
    c Use ideas about particles to explain your answer.
4 a After how many seconds was the experiment using the single lump of limestone complete?
    b How can you tell this from the graph?
5 What other factors must be kept constant during this experiment to show the effect of surface area on the rate of reaction?
6 Can you think of any sources of errors with the method used?

| Time/s | A lump of limestone | Smaller fragments of limestone |
	Volume of CO_2/cm³	Volume of CO_2/cm³
0	0	0
10	12	25
20	25	42
30	36	53
40	46	62
50	54	66
60	61	68
70	65	68
80	68	68
90	68	68
100	68	68

An increase in the surface area of a solid reactant results in an increase in the number of collisions, and this results in an increase in the number of successful collisions. Therefore, the increase in surface area of the limestone increases the rate of reaction.

In certain industries the large surface area of fine powders and dusts can be a problem. For example, there is a risk of explosion in flourmills and mines, where the large surface area of the flour or coal dust can – and has – resulted in explosions through a reaction with oxygen gas in the air when a spark has been created by machinery or the workforce (Figure 7.7). On 26 September 1988, two silos containing wheat exploded at the Jamaica Flour Mills Plant in Kingston, Jamaica, killing three workers, as a result of fine dust exploding.

▲ **Figure 7.7** The dust created by this cement plant is a potential hazard

---

Test yourself

1 What apparatus would you use to measure the rate of reaction of limestone with dilute hydrochloric acid by measuring the volume of carbon dioxide produced?

2 The following results were obtained from an experiment of the type you were asked to design in question **1**.

Time/min	0	0.5	1.0	1.5	2.0	2.5	3.0	3.5	4.0	4.5	5.0
Total volume of CO_2 gas/cm³	0	15	24	28	31	33	35	35	35	35	35

   a Plot a graph of the total volume of CO_2 against time.
   b At which point is the rate of reaction fastest?
   c What volume of CO_2 was produced after 1 minute 15 seconds?
   d How long did it take to produce 30 cm³ of CO_2?

Concentration

A yellow precipitate is produced in the reaction between sodium thiosulfate and hydrochloric acid.

sodium + hydrochloric → sodium + sulfur + sulfur + water
thiosulfate   acid   chloride   dioxide

$$Na_2S_2O_3(aq) + 2HCl(aq) \rightarrow 2NaCl(aq) + S(s) + SO_2(g) + H_2O(l)$$

The rate of this reaction can be followed by recording the time taken for a given amount of the yellow sulfur to be precipitated. This can be done by placing a conical flask containing the reaction mixture on to a cross on a piece of paper (Figure 7.8). As the precipitate of sulfur forms, the cross is obscured and finally disappears from view. The time taken for this to occur is a measure of the rate of this reaction. To obtain sufficient information about the effect of changing the concentration of the reactants, several experiments of this type must be carried out, using different concentrations of sodium thiosulfate or hydrochloric acid. If considering carrying out this practical, teachers should take care as SO_2 is released, which can cause breathing issues, especially for asthmatics.

▲ **Figure 7.8** The precipitate of sulfur obscures the cross

Some sample results of experiments of this kind have been plotted in Figure 7.9. You can see from the graph that when the most concentrated sodium thiosulfate solution was used, the reaction was at its fastest. This is shown by the shortest time taken for the cross to be obscured.

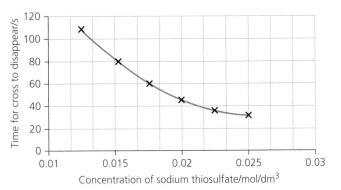

▲ **Figure 7.9** Sample data for the sodium thiosulfate/acid experiment at different concentrations of sodium thiosulfate

From the data shown in Figure 7.9 it is possible to produce a different graph which directly shows the rate of the reaction against concentration rather than the time taken for the reaction to occur against concentration. To do this, the times can be converted to a rate using:

$$\text{rate} = \frac{1}{\text{reaction time (s)}}$$

This would give the graph shown in Figure 7.10.

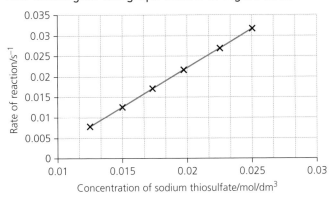

▲ **Figure 7.10** Graph to show the rate of reaction against concentration

The products of a reaction are formed as a result of the collisions between reactant particles. In a more concentrated solution there are more particles, which leads to more frequent collisions between reactant particles. The more often the particles collide, the greater the chance they have of having sufficient energy to overcome the activation energy of the reaction, and so it is more likely that a successful collision occurs. This means that the rate of a chemical reaction will increase if the concentration of reactants is increased, because there are more particles per unit volume.

Pressure of gases

In reactions involving only gases, an increase in the overall pressure at which the reaction is carried out increases the rate of the reaction. The increase in pressure results in the gas particles being pushed closer together, increasing the number of particles per unit volume. This means that they collide more frequently and so the rate of reaction increases.

Temperature

Why do you think food is stored in a refrigerator? The reason is that the rate of decay of food is slower at lower temperatures. This is a general feature of the majority of chemical processes.

The reaction between sodium thiosulfate and hydrochloric acid can be used to study the effect of temperature on the rate of a reaction. Figure 7.11 shows some sample results of experiments with sodium thiosulfate and hydrochloric acid (at fixed concentrations) carried out at different temperatures. You can see from the graph that the rate of the reaction is fastest at high temperatures.

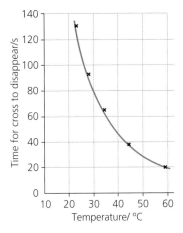

▲ **Figure 7.11** Sample data for the sodium thiosulfate/acid experiment at different temperatures

As the temperature increases, the reactant particles increase their kinetic energy. Some of the particles will have an energy greater than the activation energy, E_a, and they move faster. The faster movement results in more frequent collisions between the particles. Some of the extra collisions that result from the temperature increase will be successful collisions. This causes the reaction rate to increase.

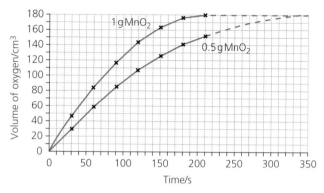

▲ **Figure 7.12** Sample data for differing amounts of MnO_2 catalyst

> **Test yourself**
>
> 3 Explain why potatoes cooked in oil cook faster than those cooked in water.
> 4 Devise an experiment to study the effect of temperature on the reaction between magnesium and hydrochloric acid.
> 5 Explain why food cooks faster in a pressure cooker.

Catalysts

Over 90% of industrial processes use **catalysts**. A catalyst is a substance which can alter the rate of a reaction without being chemically changed itself. In the laboratory, the effect of a catalyst can be observed using the decomposition of hydrogen peroxide as an example.

$$\text{hydrogen peroxide} \rightarrow \text{water} + \text{oxygen}$$
$$2H_2O_2(aq) \rightarrow 2H_2O(l) + O_2(g)$$

The rate of decomposition at room temperature is very slow. There are substances that can speed up this reaction, one being manganese(IV) oxide. When black manganese(IV) oxide powder is added to hydrogen peroxide solution, oxygen is produced rapidly. The rate at which this occurs can be seen by measuring the volume of oxygen gas produced with time.

Some results from experiments of this type have been plotted in Figure 7.12. At the end of the reaction, the manganese(IV) oxide can be filtered off and used again. The reaction can proceed even faster by increasing the amount and surface area of the catalyst. This is because the activity of a catalyst involves its surface. Note that, in gaseous reactions, if dirt or impurities are present on the surface of the catalyst, it will not act as efficiently;

it is said to have been 'poisoned'. Therefore, the gaseous reactants must be pure.

Chemists have found that:

» a small amount of catalyst will produce a large amount of chemical change
» catalysts remain unchanged chemically after a reaction has taken place, but they can change physically. For example, a finer manganese(IV) oxide powder is left behind after the decomposition of hydrogen peroxide
» catalysts are specific to a particular chemical reaction.

Some examples of chemical processes and the catalysts used are shown in Table 7.1.

▼ **Table 7.1** Examples of catalysts

Process	Catalyst
Haber process – for the manufacture of ammonia	Iron
Contact process – for the manufacture of sulfuric acid	Vanadium(V) oxide
Oxidation of ammonia to give nitric acid	Platinum
Hydrogenation of unsaturated oils to form fats in the manufacture of margarines	Nickel

A catalyst increases the rate of a chemical reaction by providing an alternative reaction path which has a lower activation energy, E_a. A catalyst does not increase the number of collisions between the reactant particles but it causes more of the collisions to become successful collisions, so increasing the rate of the reaction.

If the activation energy is lowered by using a catalyst then, on collision, more particles will go on to produce products at a given temperature (Figure 7.13).

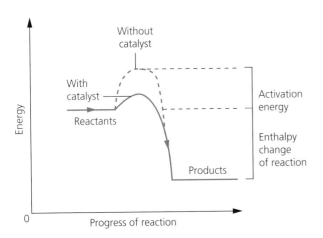

▲ **Figure 7.13** A reaction pathway diagram showing activation energy, with and without a catalyst

> **Key definition**
>
> A **catalyst** is a substance which alters the rate of a chemical reaction and is unchanged at the end of the reaction. It increases the rate of a chemical reaction by providing an alternative reaction path which has a lower activation energy, E_a.

> **Test yourself**
>
> 6 Using a catalysed reaction of your choice, devise an experiment to follow the progress of the reaction and determine how effective the catalyst is.

7.3 Enzymes

Enzymes are protein molecules produced in living cells. They are catalysts which speed up hundreds of different chemical reactions taking place inside living cells. These biological catalysts are very specific in that each chemical reaction has a different enzyme catalyst.

> **Test yourself**
>
> 7 When using biological washing powders what factors have to be taken into consideration?
> 8 Enzymes in yeast are used in the fermentation of glucose. Why, when the temperature is raised to 45°C, is very little ethanol actually produced compared with the amount formed at room temperature?

Rate of reaction is very important in the chemical industry. There are, however, many other factors which the industry needs to take into account as they manufacture chemicals such as safety and cost. In the next section we will look at the industrial manufacture of a very important chemical called ammonia.

Going further

There are literally hundreds of different kinds of enzyme. Enzymes all have an active site. The active site is a particular shape and locks into a corresponding shape in a reactant molecule. When this has happened, the enzyme can work to break up the reactant (Figure 7.14).

For example, hydrogen peroxide is produced within our bodies. However, it is extremely damaging and must be decomposed very rapidly. Catalase is the enzyme which converts hydrogen peroxide into harmless water and oxygen within our livers:

$$\text{hydrogen peroxide} \xrightarrow{\text{catalase}} \text{water} + \text{oxygen}$$

$$2H_2O_2(aq) \xrightarrow{\text{catalase}} 2H_2O(l) + O_2(g)$$

Although many chemical catalysts can work under various conditions of temperature and pressure, as well as alkalinity or acidity, biological catalysts operate only under very particular conditions. For example, they

▲ **Figure 7.14** The enzyme molecules (red, pink, green and blue) lock on exactly to a particular reactant molecule (yellow). Once the enzyme is locked on, the reactant molecule breaks up into pieces

operate over a very narrow temperature range and, if the temperature becomes too high, they become inoperative.

At temperatures above about 45°C, they denature. This means that the specific shape of the active site of the enzyme molecule changes due to the breaking of bonds. This means that the reactant molecules are no longer able to fit into the active site.

A huge multimillion-dollar industry has grown up around the use of enzymes to produce new materials. Biological washing powders (Figure 7.15) contain enzymes to break down stains such as sweat, blood and egg, and they do this at the relatively low temperature of 40°C. This reduces energy costs, because the washing water does not need to be heated as much.

▲ **Figure 7.15** These biological washing powders contain enzymes

There have been problems associated with the early biological washing powders. Some customers suffered from skin rashes, because they were allergic to the enzymes. Manufacturers usually place warnings on the packets to indicate that the powder contains enzymes which may cause skin rashes, and also advise that extra rinsing is required.

Other industrial processes also make use of enzymes.

- In the manufacture of baby foods, enzymes called proteases are used to 'pre-digest' the protein part of the baby food.
- In the production of yoghurt, enzymes are added to partially digest milk protein, allowing it to be more easily absorbed by the body.
- In cheese making, enzymes are added to coagulate milk and to separate it into solids (curds) and liquid (whey).

In industry, enzymes are used to bring about reactions at normal temperatures and pressures that would otherwise require expensive conditions and equipment. Successful processes using enzymes need to ensure that:

- the enzyme is able to function for long periods of time by optimising the environment
- the enzyme is kept in place by trapping it on the surface of an inert solid (some processes immobilise the enzymes when the process is complete)
- continuous processes occur rather than batch processes.

7.4 Reversible reactions and equilibrium

Most chemical reactions simply change reactants into products. For example, carbon reacts with oxygen to produce carbon dioxide, but the carbon dioxide produced does not easily change back to give carbon and oxygen again. There are, however, some chemical reactions which are reversible. These are reactions in which reactants form products which can then react further to produce more reactants, which reverse the process and produce the original reactants. The resulting mixture may contain reactants and products, and their proportions can be changed by altering the reaction conditions.

Some of these reactions produce important industrial chemicals, such as ammonia and sulfuric acid. We will look at the industrial production of ammonia later in this chapter but first let us look at some simpler examples of reversible reactions.

Copper(II) sulfate is available in two different forms: **hydrated** copper(II) sulfate, which is a blue solid, and **anhydrous** copper(II) sulfate which is a white solid. If hydrated copper(II) sulfate is heated, water is removed and the colour changes from blue to white, as anhydrous copper(II) sulfate is formed. If water is then added to the white solid, it changes back to a blue colour as the hydrated copper(II) sulfate is re-formed (Figure 7.16).

7.5 Ammonia – an important nitrogen-containing chemical

Nitrogen from the air is used to manufacture ammonia, a very important bulk chemical. A bulk chemical is one that is used in large quantities, across a range of uses. The major process used for making ammonia is the **Haber process**. This process was developed by the German scientist Fritz Haber in 1913. He was awarded a Nobel Prize in 1918 for his work. The process involves reacting nitrogen and hydrogen. Ammonia has many uses including the manufacture of explosives, nitric acid and fertilisers such as ammonium nitrate.

Obtaining nitrogen

The nitrogen needed in the Haber process is obtained from the **atmosphere** by fractional distillation of liquid air (Chapter 12, p. 220).

Obtaining hydrogen

The hydrogen needed in the Haber process is obtained from the reaction between methane and steam.

$$\text{methane} + \text{steam} \rightleftharpoons \text{hydrogen} + \text{carbon monoxide}$$

$$CH_4(g) + H_2O(g) \rightleftharpoons 3H_2(g) + CO(g)$$

This process is known as steam re-forming and is reversible.

In any reversible reaction, the reaction shown going from left to right in the equation is known as the forward reaction. The reaction shown going from right to left in the equation is known as the reverse reaction. Special conditions are employed to ensure that the reaction proceeds as the forward reaction (to the right in the equation), producing hydrogen and carbon monoxide. The process is carried out at a temperature of 750°C, at a pressure of 3000 kPa (30 atmospheres) with a catalyst of nickel. These conditions enable the maximum amount of hydrogen to be produced at an economic cost.

The carbon monoxide produced is then allowed to reduce some of the unreacted steam to produce more hydrogen gas.

$$\text{carbon monoxide} + \text{steam} \rightleftharpoons \text{hydrogen} + \text{carbon dioxide}$$

$$CO(g) + H_2O(g) \rightleftharpoons H_2(g) + CO_2(g)$$

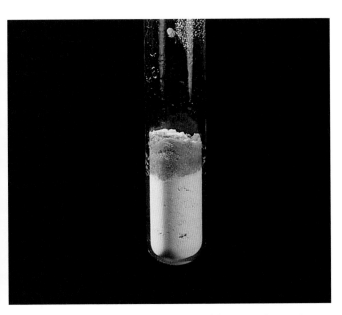

▲ **Figure 7.16** Anhydrous copper(II) sulfate changes from white to blue when water is added to it

This is an example of a reversible reaction:

$$\text{hydrated copper(II)} \rightleftharpoons \text{anhydrous copper(II)} + \text{water}$$
sulfate   sulfate

blue solid $\rightleftharpoons$ white solid

We can show this reaction as being **reversible** by using the $\rightleftharpoons$ symbol.

> **Key definition**
> Some chemical reactions are **reversible** as shown by the $\rightleftharpoons$ symbol. This means the reaction can go both ways.

Another example is cobalt(II) chloride, which is a blue solid in its anhydrous form and pink in its hydrated form. Adding water to anhydrous cobalt(II) chloride changes the colour from blue to a pink solid as it produces hydrated cobalt(II) chloride. If hydrated cobalt(II) chloride is heated the colour changes from pink back to blue.

$$\text{anhydrous cobalt(II)} + \text{water} \rightleftharpoons \text{hydrated cobalt(II)}$$
chloride   chloride

blue solid   pink solid

You will learn more about copper(II) sulfate in Chapter 8, p. 119. Both examples shown above can be used as a chemical test for the presence of water.

In summary, heating a hydrated compound will remove the water and form the anhydrous compound. Adding some water to an anhydrous compound will produce the hydrated compound.

Other ways of producing hydrogen for the Haber process do exist, usually using hydrocarbons, but steam re-forming is the most commonly used method.

> **Key definition**
>
> The source of hydrogen used in the **Haber process** is methane, and nitrogen gas is obtained from the air.

Making ammonia

In the Haber process, nitrogen and hydrogen in the correct proportions (1 : 3) are pressurised to approximately 20 000 kPa (200 atmospheres) and passed over a catalyst of freshly produced, finely divided iron at a temperature of 450°C. The reaction in the Haber process is:

nitrogen + hydrogen $\rightleftharpoons$ ammonia

$$N_2(g) \ + \ 3H_2(g) \ \rightleftharpoons 2NH_3(g); \Delta H = -92\,kJ/mol$$

The reaction is exothermic.

> **Key definition**
>
> The equation for the production of ammonia in the Haber process is $N_2 + 3H_2 \rightleftharpoons 2NH_3$.
>
> The typical conditions used in the Haber process are 450°C, 20 000 kPa and an iron catalyst.

The industrial conditions employed ensure that sufficient ammonia is produced at a fast enough rate. Under these conditions, the gas mixture leaving the reaction vessel contains about 15% ammonia, which is removed by cooling and condensing it as a liquid. The unreacted nitrogen and hydrogen are re-circulated into the reaction vessel to react together once more to produce further quantities of ammonia.

Equilibrium

The 15% of ammonia produced does not seem a great deal. The reason for this is the reversible nature of the reaction. Once the ammonia is made from nitrogen and hydrogen, it decomposes to produce nitrogen and hydrogen. There comes a point when the rate at which the nitrogen and hydrogen react to produce ammonia is equal to the rate at which the ammonia decomposes. The concentrations of nitrogen, hydrogen and ammonia in the reaction vessel do not change, provided that the reaction vessel is a closed system: one in which none of the gases can enter or leave. This situation is called

chemical **equilibrium**. Because the processes continue to happen, the equilibrium is said to be dynamic. A **reversible reaction**, in a closed system, is at equilibrium when the rate of the forward and reverse reactions are equal and the concentrations of the reactants and products remain constant. The conditions used ensure that the ammonia is made economically. Figure 7.17 shows how the percentage of ammonia produced varies with the use of different temperatures and pressures.

> **Key definition**
>
> A **reversible reaction** in a closed system is at **equilibrium** when the rate of the forward reaction is equal to the rate of the reverse reaction, and the concentrations of reactants and products are no longer changing.

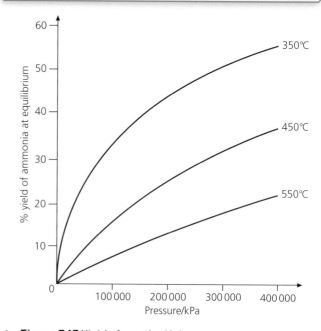

▲ **Figure 7.17** Yields from the Haber process

You will notice from Figure 7.17 that the higher the pressure and the lower the temperature used, the more ammonia is produced. In fact, the position of equilibrium is affected by each of the following:

» changing temperature
» changing pressure
» changing concentration
» using a catalyst.

Relationships such as these were initially observed by Henri Le Chatelier, a French scientist, in 1888 (Figure 7.18). He noticed that if the pressure was increased in reactions involving gases, the reaction which produced the fewest molecules of gas was favoured. If you look at the reaction for the Haber process you will see that, going from left to right, the number of molecules of gas goes from four to two. This is why the Haber process is carried out at high pressures. We say that the position of equilibrium has been moved to the right by using the higher pressure. Le Chatelier also noticed that reactions which were exothermic produced more products if the temperature was low. Using a lower temperature favours the exothermic process. In the Haber process this would move the position of equilibrium to the right. Indeed, if the Haber process is carried out at room temperature, you get a higher percentage of ammonia. However, in practice the rate of the reaction is lowered too much and the ammonia is not produced quickly enough for the process to be economical. An optimum temperature is used to produce enough ammonia at an acceptable rate.

A catalyst is used in almost all industrial processes to produce the product more quickly, and in the Haber process an iron catalyst is used. The catalyst increases the rate of both the forward and reverse reactions to the same extent, so the position is not affected: its use simply means the product is produced more quickly.

The concentration of nitrogen is often increased as this will also favour the forward reaction, increasing the yield of ammonia.

It should be noted, however, that the high pressure used is very expensive and is a major safety concern so alternative, less expensive routes involving biotechnology are currently being sought.

Worldwide, in excess of 175 million tonnes of ammonia are produced by the Haber process each year.

▲ **Figure 7.18** Henri Le Chatelier (1850–1936)

> ## Test yourself
>
> 9 What problems do the builders of a chemical plant designed to produce ammonia have to consider when they start to build such a plant?
> 10 What problems are associated with building a plant which uses such high pressures as those required in the Haber process?

7.6 Industrial manufacture of sulfuric acid – the Contact process

Sulfuric acid is probably the most important industrial chemical, and the quantity of it produced by a country has been linked with the economic stability of the country. Many million tonnes of sulfuric acid are produced worldwide each year. It is used mainly as the raw material for the production of substances such as detergents, paints, fertilisers and organic compounds including plastics.

The process by which sulfuric acid is produced is known as the **Contact process.**

The process has the following stages.

Sulfur dioxide is first produced by burning sulfur with an excess of oxygen in the air.

sulfur + oxygen → sulfur dioxide

$$S(s)\ +\ O_2(g)\ \rightarrow\quad SO_2(g)$$

The sulfur dioxide can also be obtained by roasting sulfide ores such as zinc sulfide in air.

The sulfur dioxide and oxygen gases are then heated to a temperature of approximately 450°C and fed into a reaction vessel, where they are passed over a catalyst of vanadium(V) oxide (V_2O_5). This catalyses the reaction between sulfur dioxide and oxygen to produce sulfur trioxide (sulfur(VI) oxide, SO_3).

sulfur dioxide + oxygen ⇌ sulfur trioxide

$$2SO_2(g)\ +\ O_2(g)\ \rightleftharpoons\quad 2SO_3(g);$$
$$\Delta H = -197\ \text{kJ/mol}$$

This reaction is reversible and so the ideas of Le Chatelier described on p. 106 can be used to increase the proportion of sulfur trioxide in the equilibrium mixture. The forward reaction is exothermic and so would be favoured by low temperatures. The temperature of 450°C is an optimum temperature which produces sufficient sulfur trioxide at an economical rate.

Since the reaction from left to right is also accompanied by a decrease in the number of molecules of gas, it will be favoured by a high pressure. In reality, the process is run at 200 kPa (2 atmospheres) of pressure. Under these conditions, about 96% of the sulfur dioxide and oxygen are converted into sulfur trioxide. The energy produced by this reaction is used to heat the incoming gases, thereby saving money.

The sulfur trioxide produced can be reacted with water to produce sulfuric acid.

sulfur trioxide + water → sulfuric acid

$$SO_3(g)\quad + H_2O(l) \rightarrow\quad H_2SO_4(l)$$

Key definition

The equation for the production of sulfur trioxide in the **Contact process** is $2SO_2 + O_2 \rightleftharpoons 2SO_3$

The typical conditions used in the Contact process are 450°C, 200 kPa and vanadium(V) oxide catalyst.

Burning sulfur or sulfide ores in air is the source of sulfur dioxide and air is the source of oxygen in the Contact process.

▶ Test yourself

11 Produce a flow diagram to show the different processes which occur during the production of sulfuric acid by the Contact process. Write balanced chemical equations showing the processes which occur at the different stages, giving the essential raw materials and conditions used.

12 Both the following reactions are reversible:
   (i) $X_2(g) + O_2(g) \rightleftharpoons 2XO(g)$
   (ii) $2XO(g) + O_2(g) \rightleftharpoons 2XO_2(g)$
   Suggest a reason why an increase in pressure:
   a does not favour reaction (i)
   b increases the amount of XO_2 produced in reaction (ii).

Revision checklist

After studying Chapter 7 you should be able to:
- ✓ Identify a chemical and physical change.
- ✓ Describe and explain the effect of changes in concentration, pressure, surface area and temperature, and the use of a catalyst on the rate of a reaction.
- ✓ Describe and evaluate different methods of investigating rates of a reaction.
- ✓ Interpret data and graphs obtained from rate experiments.
- ✓ Describe the collision theory in terms of the number of particles per unit volume and frequency of the collisions between the particles.
- ✓ Explain that increasing the concentration of a solution, or the pressure of a gas, increases the number of particles per unit volume, which leads to more frequent collisions between them.
- ✓ Explain that increasing the temperature leads to the particles having more kinetic energy, making them accelerate and collide more frequently, causing more of the collisions to be successful.

- ✓ State that a catalyst increases the rate of a reaction and is unchanged at the end of the reaction.
- ✓ Explain that a catalyst lowers the activation energy, E_a, of a reaction which means that more of the collisions are successful.
- ✓ Use the $\rightleftharpoons$ symbol to show a reversible reaction.
- ✓ Explain when a chemical equilibrium occurs.
- ✓ Describe the effects of changing the conditions for a reversible reaction for hydrated and anhydrous compounds.
- ✓ Predict and explain how changing the conditions for a reversible reaction will alter the position of equilibrium.
- ✓ State the equation for the Haber process and the sources of the nitrogen and hydrogen needed.
- ✓ State and explain the typical conditions used in the Haber process.
- ✓ State the equation for the Contact process and the sources of the sulfur dioxide and oxygen needed.
- ✓ State and explain the typical conditions used in the Contact process.

Exam-style questions

1 Explain the following statements.
   a A car exhaust pipe will rust much faster if the car is in constant use. [2]
   b Vegetables cook faster when they are chopped up. [1]
   c Industrial processes become more economically viable if a catalyst can be found for the reactions involved. [2]
   d In fireworks it is usual for the ingredients to be powdered. [2]
   e Tomatoes ripen faster in a glasshouse. [2]
   f The reaction between zinc and dilute hydrochloric acid is slower than the reaction between zinc and concentrated hydrochloric acid. [2]

2 A student performed two experiments to establish how effective manganese(IV) oxide was as a catalyst for the decomposition of hydrogen peroxide. The results below were obtained by carrying out these experiments with two different quantities of manganese(IV) oxide. The volume of the gas produced was recorded against time.

Time/s	0	30	60	90	120	150	180	210
Volume for 0.3 g/cm³	0	29	55	79	98	118	133	146
Volume for 0.5 g/cm³	0	45	84	118	145	162	174	182

   a Draw a diagram of the apparatus you could use to carry out these experiments. [3]
   b Plot a graph of the results. [3]
   c Is the manganese(IV) oxide acting as a catalyst in this reaction? Explain your answer. [2]
   d i At which stage does the reaction proceed most quickly? [1]
     ii How can you tell this from your graph? [1]
     iii In terms of particles, explain why the reaction is quickest at the point you have chosen in i. [2]
   e Why does the slope of the graph become less steep as the reaction proceeds? [2]
   f What volume of gas has been produced when using 0.3 g of manganese(IV) oxide after 50 s? [1]

   g How long did it take for 60 cm³ of gas to be produced when the experiment was carried out using 0.5 g of the manganese(IV) oxide? [1]
   h Write a balanced chemical equation for the decomposition of hydrogen peroxide. [1]

3 A flask containing dilute hydrochloric acid was placed on a digital balance. An excess of limestone chippings was added to this acid, a plug of cotton wool was placed in the neck of the flask and the initial mass was recorded. The mass of the apparatus was recorded every two minutes. At the end of the experiment, the loss in mass of the apparatus was calculated and the following results were obtained.

Time/min	0	2	4	6	8	10	12	14	16
Loss in mass/g	0	2.1	3.0	3.1	3.6	3.8	4.0	4.0	4.0

   a Plot the results of the experiment. [3]
   b Which of the results would appear to be incorrect? Explain your answer. [2]
   c Write a balanced chemical equation to represent the reaction taking place. [1]
   d Why did the mass of the flask and its contents decrease? [1]
   e Why was the plug of cotton wool used? [1]
   f How does the rate of reaction change during this reaction? Explain this using particle theory. [2]
   g How long did the reaction last? [1]
   h How long did it take for half of the reaction to occur? [1]

4 a What is a catalyst? [2]
   b List the properties of catalysts. [3]
   c Name the catalyst used in the following processes:
     i the Haber process [1]
     ii the hydrogenation of unsaturated fats. [1]
   d Which series of metallic elements in the Periodic Table (p. 130) do the catalysts you have named in c belong to? [1]
   e What are the conditions used in the industrial processes named in c? [2]

5 Suggest practical methods by which the rate of reaction can be investigated in each of the following cases:
  a magnesium reacting with hydrochloric acid [2]
  b nitrogen monoxide reacting with oxygen. [2]
6 Ammonia gas is made industrially by the Haber process, which involves the reaction between the gases nitrogen and hydrogen. The amount of ammonia gas produced from this reaction is affected by both the temperature and the pressure at which the process is run. The graph shows how the amount of ammonia produced from the reaction changes with both temperature and pressure. The percentage yield of ammonia indicates the percentage of the nitrogen and hydrogen gases that are actually changed into ammonia gas.

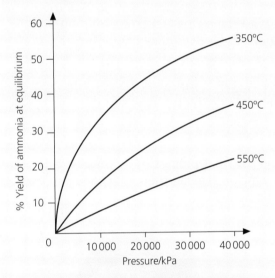

  a i Write a word and balanced chemical equation for the reversible reaction between nitrogen and hydrogen to produce ammonia using the Haber process. [2]

  ii What is the maximum mass of ammonia gas, in kg, that could be produced from 2.4×10^4 dm^3 of nitrogen gas?
  iii If the amount obtained was 13.7 kg, what was the percentage yield?
  b What is meant by the term 'reversible reaction'? [2]
  c Use the graphs to say whether more ammonia is produced at:
  i higher or lower temperatures [1]
  ii higher or lower pressures. [1]
  d i Is the forward reaction endothermic or exothermic? Explain how you decided this.
  ii Use the bond energy values in the table below to calculate the enthalpy change, ΔH, for the forward reaction.

Bond	Bond energy/kJ/mole
N≡N	946
H–H	436
N–H	390

  Did your value for ΔH match with your answer to part **i**?
  e What is the percentage yield of ammonia if the conditions used to run the process are:
  i a temperature of 350°C and a pressure of 10 000 kPa? [1]
  ii a temperature of 550°C and a pressure of 35 000 kPa? [1]
  f The conditions in industry for the production of ammonia are commonly of the order of 20 000 kPa and 450°C. What is the percentage yield of ammonia using these conditions? [1]
  g Why does industry use the conditions stated in part **f** if it is possible to obtain a higher yield of ammonia using different conditions? [2]

Acids, bases and salts

In this chapter you will look at acids and bases, their properties and their reactions. You will look at the use of indicators to tell whether a solution is acidic or alkaline, and you will be introduced to indicators other than universal indicator solution, which you have probably already met.

The reactions we will look at in this chapter produce ionic solids called salts. Salts are very important to our lives: for example, ammonium nitrate is the most used fertiliser on our planet and magnesium sulfate is used in medicines. This chapter will guide you through different methods of making salts and which reactants you would use. For example, how do you make ammonium nitrate? Which reactants will you need and which of the different methods you will read about should you use?

By the end of this chapter you should be able to answer these questions. You will also become more confident in writing chemical equations and using these equations to work out the quantities of materials you need to start with and how much of the product you should make.

8.1 Acids and alkalis

All the substances shown in Figure 8.1 contain an **acid** of one sort or another. Acids are certainly all around us. What properties do these substances have which make you think that they are acids or contain acids?

▲ **Figure 8.1** What do all these foods and drinks have in common?

Some common alkaline substances are shown in Figure 8.2.

▲ **Figure 8.2** Some common alkaline substances

It would be too dangerous to taste a liquid to find out if it was acidic. Chemists use substances called **indicators** which change colour when they are added to acids or alkalis. Many indicators are dyes which have been extracted from natural sources, such as litmus.

Methyl orange, thymolphthalein and blue and red litmus are all indicators used in titrations (see p. 124). Table 8.1 shows the colours they turn in acids and alkalis.

▼ **Table 8.1** Indicators and their colours in acid and alkaline solution

Indicator	Colour in acid solution	Colour in alkaline solution
Blue litmus	Red	Blue
Methyl orange	Pink	Yellow
Thymolphthalein	Colourless	Blue
Red litmus	Red	Blue

These indicators tell chemists whether a substance is acid or alkaline (Figure 8.3). To obtain an idea of how acidic or alkaline a substance is, we use another indicator known as a universal indicator. This indicator is a mixture of many other indicators. The colour shown by this indicator can be matched against a **pH scale**. The pH scale was developed by a Scandinavian chemist called Søren Sørenson. The pH scale runs from below 0 to 14. A substance with a pH of less than 7 is an acid. One with a pH of greater than 7 is alkaline. One with a pH of 7 is said to be neither acid nor alkaline: it is neutral. Water is the most common example of a neutral substance. Figure 8.4 shows the universal indicator colour range along with everyday substances with their particular pH values.

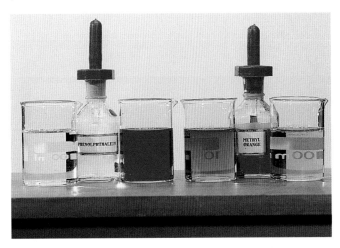

▲ **Figure 8.3** Indicators tell you if a substance is acid or alkaline

113

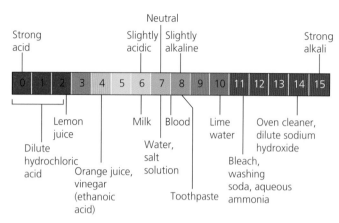

Neutral
Strong acid | Slightly acidic | Slightly alkaline | Strong alkali

0 1 2 3 4 5 6 7 8 9 10 11 12 13 14 15

Lemon juice

Dilute hydrochloric acid

Orange juice, vinegar (ethanoic acid)

Milk

Water, salt solution

Blood

Toothpaste

Lime water

Bleach, washing soda, aqueous ammonia

Oven cleaner, dilute sodium hydroxide

a The pH scale

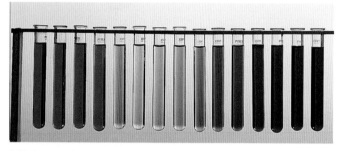

b Universal indicator in solution, showing the colour range

▲ **Figure 8.4**

Using universal indicator allows us to determine how concentrated one acid solution is compared to another. The redder the colour, the more acidic the solution is. Solutions which are very acidic contain higher concentrations of hydrogen ions ($H^+(aq)$).

Another way in which the pH of a substance can be measured is by using a pH meter (Figure 8.5). The pH electrode is placed into the solution and a pH reading is given on the digital display.

▲ **Figure 8.5** A digital pH meter

The Brønsted–Lowry theory

This theory defined:

» an **acid** as an H^+ ion (or proton) donor
» a **base** as an H^+ ion (or proton) acceptor.

> **Key definitions**
>
> **Acids** are proton donors.
>
> **Bases** are proton acceptors.

The theory explains why a pure acid behaves differently from its aqueous solution, since for an acid to behave as an H^+ ion donor it must have another substance present to accept the H^+ ion. So the water, in the aqueous acid solution, is behaving as a Brønsted–Lowry base and accepting an H^+ ion. Generally:

$$HA(aq) + H_2O(l) \rightarrow H_3O^+(aq) + A^-(aq)$$

$$\text{acid} \qquad \text{base}$$

If a substance can behave both as a proton acceptor (base) and a proton donor (acid) then it is called **amphoteric**, for example aluminium oxide, Al_2O_3, and zinc oxide, ZnO.

The relative strengths of acids and bases

The relative strength of an acid is found by comparing one acid with another. The strength of any acid depends upon how many molecules dissociate (or ionise) when the acid is dissolved in water. The relative strength of a base is found by comparing one base with another and is again dependent upon the dissociation of the base in aqueous solution.

Strong and weak acids

A typical strong acid is hydrochloric acid. It is formed by dissolving hydrogen chloride gas in water. In hydrochloric acid, the ions formed separate completely.

hydrogen chloride $\xrightarrow{\text{water}}$ hydrogen + chloride
ions   ions

$HCl(g) \xrightarrow{\text{water}} H^+(aq) + Cl^-(aq)$

> **Key definition**
>
> A **strong acid** is an acid that is completely dissociated in aqueous solution. Hydrochloric acid is a strong acid: $HCl(aq) \rightarrow H^+(aq) + Cl^-(aq)$

For hydrochloric acid, *all* the hydrogen chloride molecules break up to form H^+ ions and Cl^- ions. Any acid that behaves in this way is termed a **strong acid**. Both sulfuric acid and nitric acid also behave in this way and are therefore also termed strong acids.

One mole of sulfuric acid completely dissociates to give 2 moles of hydrogen ions and 1 mole of sulfate ions:

$$H_2SO_4(aq) \rightarrow 2H^+(aq) + SO_4^{2-}(aq)$$

All these acids have a high concentration of hydrogen ions in solution ($H^+(aq)$) and have a low pH. Their solutions are good conductors of electricity and they react quickly with metals, bases and metal carbonates.

When strong acids are neutralised by strong alkalis, the following reaction takes place between hydrogen ions and hydroxide ions.

$$H^+(aq) + OH^-(aq) \rightarrow H_2O(l)$$

A **weak acid**, such as ethanoic acid, which is found in vinegar, produces few hydrogen ions when it dissolves in water compared with a strong acid of the same concentration. It is only partially dissociated. Its solution has a higher pH than a strong acid, but still less than 7.

ethanoic acid $\xrightleftharpoons{\text{water}}$ hydrogen ions + ethanoate ions

$$CH_3COOH(aq) \xrightleftharpoons{\text{water}} H^+(aq) + CH_3COO^-(aq)$$

> **Key definition**
>
> A **weak acid** is an acid that is partially dissociated in aqueous solution. Ethanoic acid is a weak acid: $CH_3COOH(aq) \rightleftharpoons CH_3COO^-(aq) + H^+(aq)$

The $\rightleftharpoons$ sign means that the reaction is reversible. This means that if the ethanoic acid molecule breaks down to give hydrogen ions and ethanoate ions then they will react together to re-form the ethanoic acid molecule. The fact that fewer ethanoic acid molecules dissociate compared with a strong acid, and that the reaction is reversible, means that few hydrogen ions are present in the solution. Other examples of weak acids are citric acid, found in oranges and lemons; carbonic acid, found in soft drinks; sulfurous acid (acid rain) (Figure 8.6) and ascorbic acid (vitamin C).

▲ **Figure 8.6** Sulfurous acid is found in acid rain. It is a weak acid and is oxidised to sulfuric acid (a strong acid). Acid rain damages the environment quite badly

Solutions of weak acids are poorer conductors of electricity and have slower reactions with metals, bases and metal carbonates.

All acids when in aqueous solution produce hydrogen ions, $H^+(aq)$. To say an acid is a strong acid does not mean it is concentrated. The *strength* of an acid tells you how easily it dissociates (ionises) to produce hydrogen ions. The *concentration* of an acid indicates the proportions of water and acid present in aqueous solution. It is important to emphasise that a strong acid is still a strong acid even when it is in dilute solution and a weak acid is still a weak acid even if it is concentrated.

Neutralisation reactions

A common situation involving neutralisation of an acid is when you suffer from indigestion. This is caused by acid in your stomach irritating the stomach lining or throat. Normally it is treated by taking an indigestion remedy containing a substance which will react with and neutralise the acid.

In the laboratory, if you wish to neutralise a common acid such as hydrochloric acid, you can use an alkali such as sodium hydroxide. If the pH of the acid is measured when some sodium hydroxide solution is added to it, the pH increases. If equal volumes of the same concentration of hydrochloric acid and sodium hydroxide are added to one another, the resulting solution is found to have a pH of 7. The acid has undergone a **neutralisation reaction** and a neutral solution has been formed.

$$\text{hydrochloric acid} + \text{sodium hydroxide} \rightarrow \text{sodium chloride} + \text{water}$$

$$HCl(aq) + NaOH(aq) \rightarrow NaCl(aq) + H_2O(l)$$

> **Key definition**
>
> A **neutralisation reaction** occurs between an acid and a base to produce water; $H^+(aq) + OH^-(aq) \rightarrow H_2O(l)$.

As we have shown, when both hydrochloric acid and sodium hydroxide dissolve in water, the ions separate completely. We may therefore write:

$$H^+(aq)Cl^-(aq) + Na^+(aq)OH^-(aq) \rightarrow Na^+(aq) + Cl^-(aq) + H_2O(l)$$

You will notice that certain ions are unchanged on either side of the equation. They are called spectator ions and are usually taken out of the equation. The equation now becomes:

$$H^+(aq) + OH^-(aq) \rightarrow H_2O(l)$$

This type of equation is known as an **ionic equation**. The reaction between any acid and any alkali in aqueous solution produces water and can be summarised by this ionic equation. It shows the ion which causes acidity ($H^+(aq)$) reacting with the ion which causes alkalinity ($OH^-(aq)$) to produce neutral water ($H_2O(l)$). Aqueous solutions of acids contain H^+ ions and aqueous solutions of alkalis contain OH^- ions.

8.2 Formation of salts

In the example on this page, sodium chloride was produced as part of the neutralisation reaction. Compounds formed in this way are known as salts. A salt is a compound that has been formed when all the hydrogen ions of an acid have been replaced by metal ions or by the ammonium ion (NH_4^+).

Salts can be classified as those which are soluble in water or those which are insoluble in water. The general solubility rules for salts are:

» all sodium, potassium and ammonium salts are soluble
» all nitrates are soluble
» all chlorides are soluble, except lead and silver
» all sulfates are soluble, except barium, calcium and lead
» all carbonates and hydroxides are insoluble, except sodium, potassium and ammonium.

If the acid being neutralised is hydrochloric acid, salts called chlorides are formed. Other types of salts can be formed with other acids. A summary of the different types of salt along with the acid they are formed from is shown in Table 8.2.

▼ **Table 8.2** Types of salt and the acids they are formed from

Acid	Type of salt	Example
Carbonic acid	Carbonates	Sodium carbonate (Na_2CO_3)
Ethanoic acid	Ethanoates	Sodium ethanoate (CH_3COONa)
Hydrochloric acid	Chlorides	Potassium chloride (KCl)
Nitric acid	Nitrates	Potassium nitrate (KNO_3)
Sulfuric acid	Sulfates	Sodium sulfate (Na_2SO_4)

Types of oxides

There are three different types of oxides. Non-metal oxides, such as sulfur dioxide, SO_2, and carbon dioxide are acidic. In aqueous solution they produce aqueous hydrogen ions, $H^+(aq)$. Metal oxides, however, are bases. If these oxides are soluble they will dissolve in water to produce aqueous hydroxide ions, $OH^-(aq)$. For example, copper(II) oxide, CuO and calcium oxide, CaO.

The third type of oxides, and the rarest, are those described as **amphoteric oxides**. These oxides react with both acids and alkalis to produce salts. Examples of amphoteric oxides are zinc oxide, ZnO, and aluminium oxide, Al_2O_3.

8.3 Methods of preparing soluble salts

There are four general methods of preparing soluble salts. These involve the reaction of an acid with:

» excess metal
» excess insoluble carbonate
» excess insoluble base
» an alkali by titration.

Acid + metal

acid + metal → salt + hydrogen

This method can only be used with the less reactive metals such as aluminium. It would be very dangerous to use a reactive metal, such as sodium, in this type of reaction. The metals usually used in this method of salt preparation are the **MAZIT** metals, that is, **m**agnesium, **a**luminium, **z**inc, **i**ron and **t**in. A typical experimental method is given below.

Excess magnesium ribbon is added to dilute hydrochloric acid. By using an excess of magnesium ribbon we are making sure that all of the acid has reacted and some magnesium is left at the end of the reaction. During this addition an effervescence is observed due to the production of hydrogen gas. In this reaction the hydrogen ions from the hydrochloric acid gain electrons from the metal atoms as the reaction proceeds.

hydrogen ions + electrons → hydrogen gas
(from metal)

$$2H^+ \quad + \quad 2e^- \quad \rightarrow \quad H_2(g)$$

magnesium + hydrochloric → magnesium + hydrogen
acid   chloride

$$Mg(s) \quad + \quad 2HCl(aq) \quad \rightarrow \quad MgCl_2(aq) \quad + \quad H_2(g)$$

▲ **Figure 8.7** The excess magnesium is filtered in this way

The excess magnesium is removed by filtration (Figure 8.7).

The magnesium chloride solution is evaporated slowly to form a **saturated** solution of the salt (Figure 8.8).

The hot concentrated magnesium chloride solution produced is tested by dipping a cold glass rod into it. If salt crystals form at the end of the rod the solution is ready to crystallise and is left to cool. Any crystals produced on cooling are filtered and dried between clean tissues.

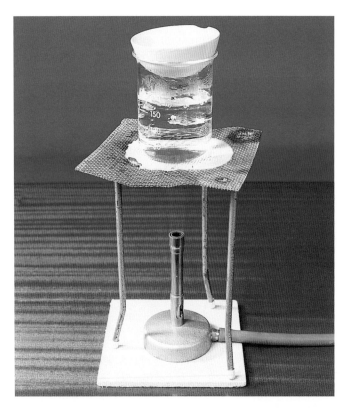

▲ **Figure 8.8** The solution of magnesium nitrate is concentrated by slow evaporation

Acid + carbonate

acid + carbonate → salt + water + carbon dioxide

This method can be used with any metal carbonate and any acid, providing the salt produced is soluble. The typical experimental procedure is similar to that carried out for an acid and a metal. For example, copper(II) carbonate would be added in excess to dilute nitric acid. Effervescence would be observed due to the production of carbon dioxide.

| copper(II) carbonate | + | nitric acid | → | copper(II) nitrate | + | carbon dioxide | + water |

$$CuCO_3(s) + 2HNO_3(aq) → Cu(NO_3)_2(aq) + CO_2(g) + H_2O(l)$$

Metal carbonates contain carbonate ions, CO_3^{2-}. In this reaction the carbonate ions react with the hydrogen ions in the acid.

| carbonate ions | + | hydrogen ions | → | carbon dioxide | + water |

$$CO_3^{2-}(aq) + 2H^+(aq) → CO_2(g) + H_2O(l)$$

When excess carbonate has been added you will see it collect at the bottom of the beaker and the effervescence will stop.

The excess copper(II) carbonate can be filtered off and the copper(II) nitrate solution is evaporated slowly to form a saturated solution of the salt.

The hot concentrated copper(II) nitrate solution produced is tested by dipping a cold glass rod into it. If salt crystals form at the end of the rod, the solution is ready to crystallise and is left to cool. Any crystals produced on cooling are filtered and dried between clean tissues.

Acid + alkali (soluble base)

acid + alkali → salt + water

Titration is generally used for preparing the salts of very reactive metals, such as potassium or sodium. It would be too dangerous to add the metal directly to the acid. In this case, we solve the problem indirectly and use an alkali which contains the reactive metal whose salt we wish to prepare.

Most metal oxides and hydroxides are insoluble bases: they are said to be basic. A few metal oxides and hydroxides that do dissolve in water to produce $OH^-(aq)$ ions are known as **alkalis**, or **soluble bases**. If the metal oxide or hydroxide does not dissolve in water, it is known as an **insoluble base**.

> **Key definitions**
>
> **Alkalis** are soluble **bases**.
>
> Aqueous solutions of alkalis contain OH^- ions.

A **base** is a substance which neutralises an acid, producing a salt and water as the only products. If the base is soluble, the term 'alkali' can be used, but there are several bases which are insoluble. It is also a substance which accepts a hydrogen ion (see p. 119). In general, most metal oxides and hydroxides (as well as ammonia solution) are bases. Some examples of soluble and insoluble bases are shown in Table 8.3. Salts can be formed by this method only if the base is soluble.

▼ **Table 8.3** Examples of soluble and insoluble bases

Soluble bases (alkalis)	Insoluble bases
Sodium hydroxide (NaOH)	Iron(III) oxide (Fe_2O_3)
Potassium hydroxide (KOH)	Copper(II) oxide (CuO)
Calcium hydroxide (Ca(OH)$_2$)	Lead(II) oxide (PbO)
Ammonia solution (NH$_3$(aq))	Magnesium oxide (MgO)

In this neutralisation reaction, both reactants are in solution, so a special technique called titration is required. Acid is slowly and carefully added to a measured volume of alkali using a **burette** (Figure 8.9) until the indicator, for example thymolphthalein, changes colour.

An indicator is used to show when the alkali has been neutralised completely by the acid. This is called the **end-point**. Once you know where the end-point is, you can add the same volume of acid to the measured volume of alkali but this time without the indicator.

▲ **Figure 8.9** The acid is added to the alkali until the indicator just changes colour

For example, the salt sodium chloride can be made by reacting sodium hydroxide with hydrochloric acid using this method.

$$\text{hydrochloric acid} + \text{sodium hydroxide} \rightarrow \text{sodium chloride} + \text{water}$$

$$HCl(aq) + NaOH(aq) \rightarrow NaCl(aq) + H_2O(l)$$

As previously discussed on p. 116, this reaction can best be described by the ionic equation:

$$H^+(aq) + OH^-(aq) \rightarrow H_2O(l)$$

The sodium chloride solution which is produced can then be evaporated slowly to obtain the salt.

The hot concentrated sodium chloride solution produced is tested by dipping a cold glass rod into it. If salt crystals form at the end of the rod, the solution is ready to crystallise and is left to cool. Any crystals produced on cooling are filtered and dried between clean tissues.

Acid + insoluble base

$$\text{acid} + \text{base} \rightarrow \text{salt} + \text{water}$$

This method can be used to prepare a salt of an unreactive metal, such as lead or copper. In these cases it is not possible to use a direct reaction of the metal with an acid so the acid is neutralised using the particular metal oxide (Figure 8.10).

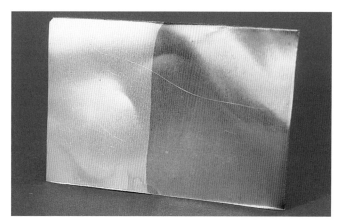

▲ **Figure 8.10** Citric acid has been used on the right-hand side of this piece of copper metal to remove the oxide coating on its surface. The acid reacts with the oxide coating to form water, leaving shiny copper metal

The method is generally the same as that for a metal carbonate and an acid, though some warming of the reactants may be necessary. An example of such a reaction is the neutralisation of sulfuric acid by copper(II) oxide to produce copper(II) sulfate (Figure 8.11).

▲ **Figure 8.11** After slow evaporation to concentrate the solution, the solution is left to crystallise. Crystals of copper(II) sulfate are produced

$$\text{sulfuric acid} + \text{copper(II) oxide} \rightarrow \text{copper(II) sulfate} + \text{water}$$

$$H_2SO_4(aq) + CuO(s) \rightarrow CuSO_4(aq) + H_2O(l)$$

Metal oxides contain the oxide ion, O^{2-}. The ionic equation for this reaction is therefore:

$$2H^+(aq) + O^{2-}(s) \rightarrow H_2O(l)$$

or

$$CuO(s) + 2H^+(aq) \rightarrow Cu^{2+}(aq) + H_2O(l)$$

Reaction of a base with ammonium salts

Small quantities of ammonia gas, NH_3, can be produced by heating any ammonium salt, such as ammonium chloride, with a base, such as calcium hydroxide. This reaction can be used to make ammonia in a laboratory.

calcium + ammonium → calcium + water + ammonia
hydroxide   chloride   chloride

$$Ca(OH)_2(s) + \quad 2NH_4Cl \quad \rightarrow \quad CaCl_2 \; + \; H_2O \; + \; 2NH_3$$

The ammonia produced can be detected as being formed by its pungent odour and by turning damp red litmus blue.

8.4 Preparing insoluble salts

The methods described above can be used to make a salt that is soluble in water. If a salt that is insoluble in water needs to be prepared, a different technique is needed.

An insoluble salt, such as barium sulfate, can be made by precipitation. In this case, solutions of the two chosen soluble salts are mixed (Figure 8.12). To produce barium sulfate, barium chloride and sodium sulfate can be used. The barium sulfate precipitate can be filtered off, washed with distilled water and dried. The reaction that has occurred is:

barium   +   sodium   →   barium + sodium
chloride   sulfate   sulfate   chloride

$$BaCl_2(aq) + Na_2SO_4(aq) \rightarrow BaSO_4(s) + 2NaCl(aq)$$

The ionic equation for this reaction is:

$$Ba^{2+}(aq) + SO_4^{2-}(aq) \rightarrow BaSO_4(s)$$

▲ **Figure 8.12** When barium chloride solution is added to sodium sulfate, a white precipitate of barium sulfate forms

This method may be summarised as follows:

soluble + soluble → insoluble + soluble
 salt   salt   salt   salt

$$(AX) + (BY) \rightarrow (BX) + (AY)$$

It should be noted that even salts like barium sulfate dissolve to a very small extent. For example, $1\,dm^3$ of water will dissolve 2.2×10^{-3} g of barium sulfate at 25°C. This substance and substances like it are said to be sparingly soluble.

8.5 Testing for different salts

Sometimes we want to analyse a salt and find out what is in it. There are simple chemical tests which allow us to identify the anion part of the salt.

Testing for a sulfate (SO_4^{2-})

You have seen that barium sulfate is an insoluble salt (p. 116). Therefore, if you take a solution of a suspected sulfate and add it to a solution of a soluble barium salt (such as barium chloride) then a white precipitate of barium sulfate will be produced.

barium ion + sulfate ion → barium sulfate

$$Ba^{2+}(aq) \; + \; SO_4^{2-}(aq) \; \rightarrow \quad BaSO_4(s)$$

A few drops of dilute hydrochloric acid are also added to this mixture. If the precipitate does not dissolve, then it is barium sulfate and the unknown salt was in fact a sulfate. If the precipitate does dissolve, then the unknown salt may have been a sulfite (containing the SO_3^{2-} ion).

Testing for a chloride (Cl^-), a bromide (Br^-) or an iodide (I^-)

Earlier in this chapter you saw that silver chloride is an insoluble salt (p. 116). Therefore, if you take a solution of a suspected chloride and add to it a small volume of dilute nitric acid, to make an aqueous acidic solution, followed by a small amount of a solution of a soluble silver salt (such as silver nitrate), a white precipitate of silver chloride will be produced.

chloride ion + silver ion → silver chloride

$$Cl^-(aq) \ + \ Ag^+(aq) \ \rightarrow \ \ AgCl(s)$$

If left to stand, the precipitate goes grey (Figure 8.13).

▲ **Figure 8.13** If left to stand the white precipitate of silver chloride goes grey. This photochemical change plays an essential part in black and white photography

In a similar way, a bromide and an iodide will react to produce either a cream precipitate of silver bromide (AgBr) or a yellow precipitate of silver iodide (AgI) (Figure 8.14).

▲ **Figure 8.14** AgCl, a white precipitate, AgBr, a cream precipitate, and AgI, a yellow precipitate

An alternative test for iodide ions is the addition of lead nitrate solution to the iodide which results in a bright yellow precipitate of lead iodide, PbI_2.

Testing for a carbonate

If a small amount of an acid is added to some of the suspected carbonate (either solid or in solution) then effervescence occurs. If it is a carbonate then carbon dioxide gas is produced, which will turn limewater 'milky' (a cloudy white precipitate of calcium carbonate forms, see Chapter 14, p. 227).

carbonate + hydrogen → carbon + water
ions   ions   dioxide

$$CO_3^{2-}(aq) \ + 2H^+(aq) \ \ \rightarrow \ CO_2(g) \ + \ H_2O(l)$$

Testing for a nitrate

By adding aqueous sodium hydroxide and then aluminium foil and warming gently, nitrates are reduced to ammonia. The ammonia can be identified using damp indicator paper, which turns blue.

$$3NO_3^-(aq) + 8Al(s) + 5OH^-(aq) + 18H_2O(l)$$
$$\downarrow$$
$$3NH_3(g) + 8[Al(OH)_4]^-(aq)$$

In the reaction the nitrate ion is reduced, as oxygen is removed from the nitrogen atom, and it gains hydrogen to form ammonia, NH_3. The gain of hydrogen is also a definition of reduction.

7 Complete the word equations and write balanced chemical equations for the following soluble salt preparations:
 a magnesium + sulfuric acid →
 b calcium carbonate + hydrochloric acid →
 c zinc oxide + hydrochloric acid →
 d potassium hydroxide + nitric acid →
 Also write ionic equations for each of the reactions.
8 Lead carbonate and lead iodide are insoluble. Which two soluble salts could you use in the preparation of each substance? Write:
 a a word equation
 b a symbol equation
 c an ionic equation
 to represent the reactions taking place.
9 An analytical chemist working for an environmental health organisation has been given a sample of water which is thought to have been contaminated by a sulfate, a carbonate and a chloride.
 a Describe how she could confirm the presence of these three types of salt by simple chemical tests.
 b Write ionic equations to help you explain what is happening during the testing process.

▼ **Table 8.4** Examples of crystal hydrates.

Salt hydrate	Formula
Cobalt(II) chloride hexahydrate	$CoCl_2.6H_2O$
Copper(II) sulfate pentahydrate	$CuSO_4.5H_2O$
Iron(II) sulfate heptahydrate	$FeSO_4.7H_2O$
Magnesium sulfate heptahydrate	$MgSO_4.7H_2O$

▲ **Figure 8.15** Hydrate crystals (left to right): cobalt nitrate, calcium nitrate and nickel sulfate (top) and manganese sulfate, copper sulfate and chromium potassium sulfate (bottom)

8.6 Water of crystallisation

Some salts, such as sodium chloride, copper carbonate and sodium nitrate, crystallise in their **anhydrous** forms (without water). However, many salts produce **hydrates** when they crystallise from solution. A hydrated substance is a salt which incorporates water into its crystal structure. This water is referred to as **water of crystallisation**. The shape of the crystal hydrate is very much dependent on the presence of water of crystallisation. Some examples of crystal hydrates are given in Table 8.4 and shown in Figure 8.15.

Key definitions

A hydrated substance, or **hydrate** is one that is chemically combined with water.

An **anhydrous** substance is one containing no water.

Water of crystallisation is the water molecules present in crystals, e.g. $CuSO_4.5H_2O$ and $CoCl_2.6H_2O$.

When many hydrates are heated, the water of crystallisation is driven away. For example, if crystals of copper(II) sulfate hydrate (blue) are heated strongly, they lose their water of crystallisation. Anhydrous copper(II) sulfate remains as a white powder:

copper(II) sulfate → anhydrous copper(II) + water
pentahydrate   sulfate

$$CuSO_4.5H_2O(s) \rightarrow \quad CuSO_4(s) \quad + 5H_2O(g)$$

When water is added to anhydrous copper(II) sulfate, the reverse process occurs. It turns blue and the pentahydrate is produced (Figure 8.16). This is an extremely exothermic process.

$$CuSO_4(s) + 5H_2O(l) \rightarrow CuSO_4.5H_2O(s)$$

▲ **Figure 8.16** Anhydrous copper(II) sulfate is a white powder which turns blue when water is added to it

Because the colour change only takes place in the presence of water, the reaction is used to test for the presence of water.

These processes give a simple example of a reversible reaction:

$$CuSO_4(s) + 5H_2O(l) \rightleftharpoons CuSO_4.5H_2O(s)$$

 Going further

Calculation of water of crystallisation

Sometimes it is necessary to work out the percentage, by mass, of water of crystallisation in a hydrated salt.

? **Worked example**

Calculate the percentage by mass of water in the salt hydrate $MgSO_4.7H_2O$. (A_r: H = 1; O = 16; Mg = 24; S = 32)

M_r for $MgSO_4.7H_2O$

$$= 24 + 32 + (4 \times 16) + (7 \times 18) = 246$$

The mass of water as a fraction of the total mass of hydrate

$$= \frac{126}{246}$$

The percentage of water present

$$= \frac{126}{246} \times 100$$

$$= 51.2\%$$

 Going further

> **Test yourself**
>
> 10 Calculate the percentage by mass of water in the following salt hydrates:
>   a $CuSO_4.5H_2O$ b $Na_2CO_3.10H_2O$
>   c $Na_2S_2O_3.5H_2O$.
>
>   (A_r: H = 1; O = 16; Na = 23; S = 32; Cu = 63.5)

 Practical skills

Preparation of potassium nitrate

Safety
- Eye protection must be worn.
- Take care using indicator solution: it can stain skin and clothing.

The apparatus below can be used to prepare potassium nitrate crystals.

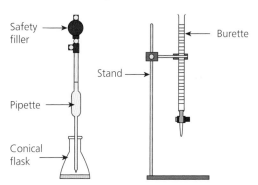

1   Clamp a burette vertically in a clamp stand and fill it up with some dilute nitric acid solution. Allow some of the acid to run through the jet of the burette. Record the initial burette reading.

2 Using a pipette and safety filler, place 25 cm³ of the potassium hydroxide solution into a 250 cm³ conical flask.
3 Put 4 drops of the indicator thymolphthalein into the potassium hydroxide in the flask. The solution will turn blue.
4 Slowly add the dilute nitric acid from the burette, swirling the flask all the time, until the blue colour disappears. Record the final burette reading and hence the amount of dilute nitric acid you needed to add to neutralise the potassium hydroxide.
5 Pour some of the solution formed in step 4 from the flask into an evaporating basin and heat until half of the solution has evaporated.
6 Set aside and allow the crystals to form slowly.

Now answer the following questions:

1 Why is it important to clamp the burette vertically?
2 Why is it important to fill up the jet of the burette before taking the initial burette reading?
3 Why is it important to use a safety filler when using the pipette?
4 How do you take the initial volume reading on the burette?
5 Why is the swirling of the flask important?
6 How could you tell that the solution has become saturated as it is evaporated?
7 In step 5, why is it important to not evaporate all the water away?
8 How can the crystals be dried?

Titration

On p. 118, you saw that it was possible to prepare a soluble salt by reacting an acid with a soluble base (alkali). The method used was that of titration. Titration can also be used to find the concentration of the alkali used. In the laboratory, the titration of hydrochloric acid with sodium hydroxide is carried out in the following way.

1 25 cm³ of sodium hydroxide solution is pipetted into a conical flask to which a few drops of thymolphthalein indicator have been added (Figure 8.17). Thymolphthalein is blue in alkaline conditions but colourless in acid.

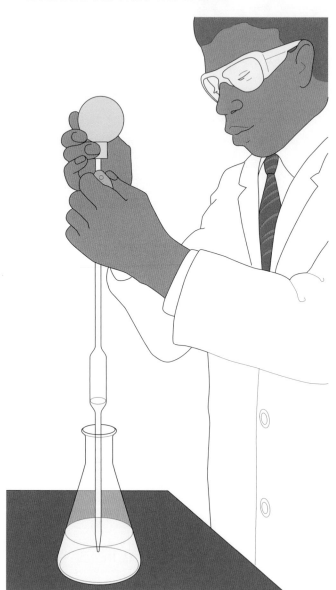

▲ **Figure 8.17** Exactly 25.0 cm³ of sodium hydroxide solution is pipetted into a conical flask

2 A 0.10 mol/dm³ solution of hydrochloric acid is placed in the burette using a filter funnel until it is filled up exactly to the zero mark (Figure 8.18).

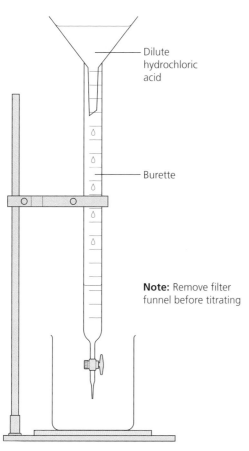

▲ **Figure 8.18** The burette is filled up to the zero mark with a 0.10 mol/dm³ solution of hydrochloric acid

3 The filter funnel is now removed.
4 The hydrochloric acid is added to the sodium hydroxide solution in small quantities – usually no more than 0.5 cm³ at a time (Figure 8.19). The contents of the flask must be swirled after each addition of acid for thorough mixing.
5 The acid is added until the alkali has been neutralised completely. This is shown by the blue colour of the indicator just disappearing.
6 The final reading on the burette at the end-point is recorded and further titrations carried out until consistent results are obtained (within 0.1 cm³ of each other). Some sample data are shown on the right.

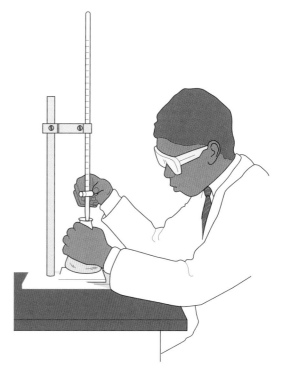

▲ **Figure 8.19** The titration is carried out accurately

Volume of sodium hydroxide solution = 25.0 cm³
Average volume of 0.10 mol/dm³ solution of hydrochloric acid added = 21.0 cm³
The neutralisation reaction which has taken place is:

hydrochloric + sodium → sodium + water
   acid   hydroxide   chloride

$$HCl(aq) + NaOH(aq) \rightarrow NaCl(aq) + H_2O(l)$$

From this equation, it can be seen that 1 mole of hydrochloric acid neutralises 1 mole of sodium hydroxide.

Now you can work out the number of moles of the acid using the formula given in Chapter 4, p. 57.

$$moles = \frac{volume}{1000} \times concentration$$

$$= 21.0 \times \frac{0.10}{10\,000}$$

$$= 2.1 \times 10^{-3}$$

number of moles of hydrochloric acid = number of moles of sodium hydroxide

Therefore, the number of moles of sodium hydroxide = 2.1×10^{-3}

2.1×10^{-3} moles of sodium hydroxide is present in 25.0 cm^3 of solution.

Therefore, in 1 cm^3 of sodium hydroxide solution we have:

$$\frac{2.1 \times 10^{-3}}{25.0} \text{ moles}$$

Therefore, in 1 dm^3 of sodium hydroxide solution we have:

$$\frac{2.1 \times 10^{-3}}{25.0} \times 1000 = 0.084 \text{ mole}$$

The concentration of sodium hydroxide solution is 0.084 mol/dm^3.

You can simplify the calculation by substituting in the following mathematical equation:

$$\frac{M_1 V_1}{M_{acid}} = \frac{M_2 V_2}{M_{alkali}}$$

where:

M_1 = concentration of the acid used

V_1 = volume of acid used (cm^3)

M_{acid} = number of moles of acid shown in the chemical equation

M_2 = concentration of the alkali used

V_2 = volume of the alkali used (cm^3)

M_{alkali} = number of moles of alkali shown in the chemical equation.

In the example:

$M_1 = 0.10$ mol/dm^3

$V_1 = 21.0$ cm^3

$M_{acid} = 1$ mole

$M_2 =$ unknown

$V_2 = 25.0$ cm^3

$M_{alkali} = 1$ mole

Substituting in the equation:

$$\frac{0.10 \times 21.0}{1} = \frac{M_2 \times 25.0}{1}$$

Rearranging:

$$M_2 = \frac{0.10 \times 21.0 \times 1}{1 \times 25.0}$$

$$M_2 = 0.084$$

The concentration of the sodium hydroxide solution is 0.084 mol/dm^3.

Another example of a titration calculation could involve a neutralisation reaction in which the ratio of the number of moles of acid to alkali is not 1 : 1. The example below shows how such a calculation could be carried out.

❓ Worked example

In a titration to find the concentration of a solution of sulfuric acid, 25 cm^3 of it were just neutralised by 20.15 cm^3 of a 0.2 mol/dm^3 solution of sodium hydroxide. What is the concentration of the sulfuric acid used?

First, write out the balanced chemical equation for the reaction taking place.

sulfuric acid + sodium hydroxide → sodium sulfate + water

$$H_2SO_4 + 2NaOH \longrightarrow Na_2SO_4 + 2H_2O$$

From this balanced equation, it can be seen that 1 mole of sulfuric acid reacts with 2 moles of sodium hydroxide.

Therefore, the number of moles of sodium hydroxide used

$$= 20.15 \times \frac{0.2}{1000} = 4.03 \times 10^{-3}$$

The number of moles of sulfuric acid which will react with 4.03×10^{-3} moles of sodium hydroxide

$$= 4.03 \times 10^{-3} \times \frac{1}{2} = 2.015 \times 10^{-3}$$

This is the number of moles of sulfuric acid present in 25 cm^3 of the solution, so the concentration of the sulfuric acid is:

$$2.015 \times 10^{-3} \times \frac{1000}{25} = 0.081 \text{ mol/}dm^3$$

Test yourself

11 24.2 cm³ of a solution containing 0.20 mol/dm³ of hydrochloric acid just neutralised 25.0 cm³ of a potassium hydroxide solution. What is the concentration of this potassium hydroxide solution?

12 22.4 cm³ of a solution containing 0.10 mol/dm³ of sulfuric acid just neutralised 25.0 cm³ of a sodium hydroxide solution. What is the concentration of this sodium hydroxide solution?

Revision checklist

After studying Chapter 8 you should be able to:
- ✓ State that acidic aqueous solutions contain OH⁻[aq] ions.
- ✓ Define an acid as a proton donor and a base as a proton donor.
- ✓ Describe how acids react with metals, bases or carbonates and how bases react with acids or ammonium salts.
- ✓ State that an alkali is a soluble base.
- ✓ Explain the difference between a concentrated and dilute acid.

- ✓ Describe the use of indicators to determine the presence of an acid or alkali and to determine their strengths.
- ✓ Explain the difference between a strong and weak acid.
- ✓ Explain what a neutralisation reaction is, and write an ionic equation.
- ✓ Classify oxides as being acidic, basic or amphoteric.
- ✓ Describe the four different methods of making soluble salts.
- ✓ State the rules of solubility.
- ✓ Explain how an insoluble salt is made.
- ✓ Define the terms 'hydrated' and 'anhydrous'.

Exam-style questions

1 Explain, with the aid of examples, what you understand by the following terms:

    a strong acid   [2]   **c** strong alkali   [2]
    b weak acid   [2]   **d** weak base.   [2]

2 a Copy out and complete the table, which covers the different methods of preparing salts.

Method of preparation	Name of salt prepared	Two substances used in the preparation
Acid + alkali	Potassium sulfate	. . . . and . . . . [2]
Acid + metal	. . . . [1]	. . . . and dilute hydrochloric acid [1]
Acid + insoluble base	Magnesium sulfate	. . . . and . . . . [2]
Acid + carbonate	Copper . . . . . [1]	. . . . and . . . . [2]
Precipitation	Lead iodide	. . . . and . . . . [2]

    b Give word and balanced chemical equations for each reaction shown in your table. Also write ionic equations where appropriate. [15]

3 Consider the following scheme.

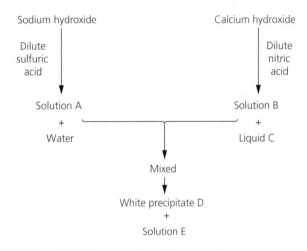

    a Give the names and formulae of substances **A** to **E**. [5]
    b Describe a test which could be used to identify the presence of water. [1]
    c Give the indicator suitable for the initial reaction between the hydroxides and the dilute acids shown. [1]

d Give balanced chemical equations for the reactions taking place in the scheme. [3]

e Give an ionic equation for the production of the white precipitate **D**. [1]

4 In a titration involving 24.0 cm³ potassium hydroxide solution against a solution of concentration 1 mol/dm³ of sulfuric acid, 28.0 cm³ of the acid was found to just neutralise the alkali completely.

a Give a word and balanced chemical equation for the reaction. [2]

b Identify a suitable indicator for the titration and state the colour change you would observe. [1]

c Calculate the concentration of the alkali in mol/dm³. [3]

d Describe a chemical test which you could use to identify the type of salt produced during the reaction. [2]

5 For each of the following, explain the term given **and** give an example.

a neutralisation [2]
b titration [3]
c soluble salt [2]
d insoluble salt. [2]

6 Copper(II) sulfate crystals exist as the *pentahydrate*, $CuSO_4.5H_2O$. It is a salt *hydrate*. If it is heated quite strongly, the *water of crystallisation* is driven off and the *anhydrous* salt remains.

a Explain the meaning of the terms shown in italics. [4]

b Describe the experiment you would carry out to collect a sample of the water given off when the salt hydrate was heated strongly. Your description should include a diagram of the apparatus used and a chemical equation to represent the process taking place. [4]

c Describe a chemical test you could carry out to show that the colourless liquid given off was water. [2]

d Describe one other test you could carry out to show that the colourless liquid obtained in this experiment was pure water. [1]

e Sometimes it is necessary to work out the percentage by mass of water of crystallisation as well as the number of moles of water present in a hydrated crystal.

i Use the information given to calculate the percentage, by mass, of water of crystallisation in a sample of hydrated magnesium sulfate. [3]

Mass of crucible = 14.20 g

Mass of crucible + hydrated $MgSO_4$ = 16.66 g

Mass after heating = 15.40 g

ii Calculate the number of moles of water of crystallisation driven off during the experiment as well as the number of moles of anhydrous salt remaining. [2]
(A_r: H = 1; O = 16; Mg = 24; S = 32)

iii Using the information you have obtained in **ii**, write down, in the form $MgSO_4.xH_2O$, the formula of hydrated magnesium sulfate. [1]

The Periodic Table

In this chapter we will look at the way in which all the chemical elements have been organised into a table that can help to predict their properties and their chemical reactions. Chemists began to try to organise the elements around 150 years ago but found it very difficult and, as new elements were discovered, they had to revise their attempts. Eventually the format of the Periodic Table which we see today was produced and, as new elements were discovered, they fitted in perfectly.

The aim of this chapter is to help you to use the Periodic Table to understand chemistry more easily. For example, if you know some of the properties of sodium, the Periodic Table can help you to predict the properties of francium.

By the end of this chapter you will have learned the trends in properties of four groups of elements and you should understand how useful the Periodic Table is to your study of chemistry.

9.1 Development of the Periodic Table

The Periodic Table is a vital tool used by chemists to predict the way in which elements react during chemical reactions. It is a method of categorising elements according to their properties. Scientists started to look for a way in which to categorise the known elements around 150 years ago.

The Periodic Table was devised in 1869 by the Russian Dmitri Mendeleev, who was the Professor of Chemistry at St Petersburg University (Figure 9.1). His periodic table was based on the chemical and physical properties of the 63 elements that had been discovered at that time.

Other scientists had attempted to categorise the known elements from the early 19th century but Mendeleev's classification proved to be the most successful.

Going further

Mendeleev arranged all the 63 known elements in order of increasing atomic weight but in such a way that elements with similar properties were in the same vertical column. He called the vertical columns **groups** and the horizontal rows **periods** (Figure 9.2). If necessary, he left gaps in the table.

As a scientific idea, Mendeleev's periodic table was tested by making predictions about elements that were unknown at that time but could possibly fill the gaps. Three of these gaps are shown by the symbols * and † in Figure 9.2. As new elements were discovered, they were

found to fit easily into the classification. For example, Mendeleev predicted the properties of the missing element 'eka-silicon' (†). He predicted the colour, density and melting point as well as its atomic weight.

In 1886 the element we now know as germanium was discovered in Germany by Clemens Winkler; its properties were almost exactly those Mendeleev had predicted. In all, Mendeleev predicted the atomic weight of ten new elements, of which seven were eventually discovered – the other three, atomic weights 45, 146 and 175, do not exist!

▲ **Figure 9.1** Dmitri Mendeleev (1834–1907)

Period	Group							
	1	2	3	4	5	6	7	8
1	H							
2	Li	Be	B	C	N	O	F	
3	Na	Mg	Al	Si	P	S	Cl	
4	K	Ca	*	Ti	V	Cr	Mn	Fe Co Ni
	Cu	Zn	*	†	As	Se	Br	

▲ **Figure 9.2** Mendeleev's periodic table. He left gaps for undiscovered elements

VIII

Period	Group																		
	I	II											III	IV	V	VI	VII		He 2, 4, Helium
1							H 1, 1, Hydrogen												
2	Li 3, 7, Lithium	Be 4, 9, Beryllium											B 5, 11, Boron	C 6, 12, Carbon	N 7, 14, Nitrogen	O 8, 16, Oxygen	F 9, 19, Fluorine		Ne 10, 20, Neon
3	Na 11, 23, Sodium	Mg 12, 24, Magnesium											Al 13, 27, Aluminium	Si 14, 28, Silicon	P 15, 31, Phosphorus	S 16, 32, Sulfur	Cl 17, 35.5, Chlorine		Ar 18, 40, Argon
4	K 19, 39, Potassium	Ca 20, 40, Calcium	Sc 21, 45, Scandium	Ti 22, 48, Titanium	V 23, 51, Vanadium	Cr 24, 52, Chromium	Mn 25, 55, Manganese	Fe 26, 56, Iron	Co 27, 59, Cobalt	Ni 28, 59, Nickel	Cu 29, 63.5, Copper	Zn 30, 65, Zinc	Ga 31, 70, Gallium	Ge 32, 73, Germanium	As 33, 75, Arsenic	Se 34, 79, Selenium	Br 35, 80, Bromine		Kr 36, 84, Krypton
5	Rb 37, 85, Rubidium	Sr 38, 88, Strontium	Y 39, 89, Yttrium	Zr 40, 91, Zirconium	Nb 41, 93, Niobium	Mo 42, 96, Molybdenum	Tc 43, 99, Technetium	Ru 44, 101, Ruthenium	Rh 45, 103, Rhodium	Pd 46, 106, Palladium	Ag 47, 108, Silver	Cd 48, 112, Cadmium	In 49, 115, Indium	Sn 50, 119, Tin	Sb 51, 122, Antimony	Te 52, 128, Tellurium	I 53, 127, Iodine		Xe 54, 131, Xenon
6	Cs 55, 133, Caesium	Ba 56, 137, Barium		Hf 72, 178.5, Hafnium	Ta 73, 181, Tantalum	W 74, 184, Tungsten	Re 75, 186, Rhenium	Os 76, 190, Osmium	Ir 77, 192, Iridium	Pt 78, 195, Platinum	Au 79, 197, Gold	Hg 80, 201, Mercury	Tl 81, 204, Thallium	Pb 82, 207, Lead	Bi 83, 209, Bismuth	Po 84, 209, Polonium	At 85, 210, Astatine		Rn 86, 222, Radon
7	Fr 87, 223, Francium	Ra 88, 226, Radium		Rf 104, 261, Rutherfordium	Db 105, 262, Dubnium	Sg 106, 263, Seaborgium	Bh 107, 262, Bohrium	Hs 108, 269, Hassium	Mt 109, 268, Meitnerium	Ds 110, 281, Darmstadtium	Rg 111, 272, Roentgenium	Cn 112, 285, Copernicium	Nh 113, 286, Nihonium	Fl 114, 285, Flerovium	Mc 115, 289, Moscovium	Lv 116, 292, Livermorium	Ts 117, 294, Tennessine		Og 118, 294, Oganesson

La 57, 139, Lanthanum	Ce 58, 140, Cerium	Pr 59, 141, Praseodymium	Nd 60, 144, Neodymium	Pm 61, 147, Promethium	Sm 62, 150, Samarium	Eu 63, 152, Europium	Gd 64, 157, Gadolinium	Tb 65, 159, Terbium	Dy 66, 162, Dysprosium	Ho 67, 165, Holmium	Er 68, 167, Erbium	Tm 69, 169, Thulium	Yb 70, 173, Ytterbium	Lu 71, 175, Lutetium
Ac 89, 227, Actinium	Th 90, 232, Thorium	Pa 91, 231, Protactinium	U 92, 238, Uranium	Np 93, 237, Neptunium	Pu 94, 244, Plutonium	Am 95, 243, Americium	Cm 96, 247, Curium	Bk 97, 247, Berkelium	Cf 98, 251, Californium	Es 99, 252, Einsteinium	Fm 100, 257, Fermium	Md 101, 258, Mendelevium	No 102, 259, Nobelium	Lr 103, 260, Lawrencium

Key

- ▢ Reactive metals
- ▢ Transition metals
- ▢ Poor metals
- ▢ Metalloids
- ▢ Non-metals
- ▢ Noble gases

▲ **Figure 9.3** The modern Periodic Table

The success of Mendeleev's predictions showed that his ideas were probably correct. His periodic table was quickly accepted by scientists as an important summary of the properties of the elements.

Mendeleev's periodic table has been modified in the light of work carried out by Ernest Rutherford and Henry Moseley. Discoveries about sub-atomic particles led them to realise that the elements should be arranged by proton number. In the modern Periodic Table, the 118 known elements are arranged in order of increasing proton number (Figure 9.3).

Those elements with similar chemical properties are found in the same columns or **groups.** There are eight groups of elements. The first column is called Group I, the second Group II, and so on up to Group VII. The final column in the Periodic Table is called Group VIII. Some of the groups have been given names.

Group I: The **alkali metals**
Group II: The **alkaline earth metals**
Group VII: The **halogens**
Group 0:   Inert gases or **noble gases**

▲ **Figure 9.4** Transition elements have a wide range of uses, both as elements and as alloys

The horizontal rows are called **periods** and these are numbered 1–7 going down the Periodic Table.

Between Groups II and III is the block of elements known as the **transition elements** (Figure 9.4).

The Periodic Table can be divided into two as shown by the bold line that starts beneath boron, in Figure 9.3. The elements to the left of this line are metals (fewer than three-quarters) and those on the right are non-metals (fewer than one-quarter). The elements which lie on this dividing line are known as **metalloids** (Figure 9.5). The metalloids behave in some ways as metals and in others as non-metals.

If you look at the properties of the elements across a period of the Periodic Table you will notice certain trends. For example, there is:

» a gradual change from metal to non-metal
» an increase in the number of electrons in the outer shell
» a change in the structure of the element, from giant metallic in the case of metals (e.g. magnesium, p. 50, Figure 3.37), through giant covalent (e.g. diamond, p. 45, Figure 3.29), to simple molecular (e.g. chlorine, p. 38, Figure 3.13).

▲ **Figure 9.5** The metalloid silicon is used to make silicon 'chips'

9.2 Electronic configuration and the Periodic Table

The number of electrons in the outer shell is discussed in Chapter 2 (p. 26). This corresponds with the number of the group in the Periodic Table in which the element is found. For example, the elements shown in Table 9.1 have one electron in their outer shell and they are all found in Group I. The elements in Group 0, however, are an exception to this rule, as they have two or eight electrons in their outer shell. The outer electrons are mainly responsible for the chemical properties of any element and, therefore, elements in the same group have similar chemical properties (Tables 9.2 and 9.3). The number of occupied shells is equal to the period number in which that element is found. For example, we know that sodium is in Group I because it has only one electron in its outer shell but we can

also say that it is in Period 3 of the Periodic Table as it has electrons in the first three electron shells.

▼ **Table 9.1** Electronic configuration of the first three elements of Group I

Element	Symbol	Proton number	Electronic configuration
Lithium	Li	3	2,1
Sodium	Na	11	2,8,1
Potassium	K	19	2,8,8,1

▼ **Table 9.2** Electronic configuration of the first three elements of Group II

Element	Symbol	Proton number	Electronic configuration
Beryllium	Be	4	2,2
Magnesium	Mg	12	2,8,2
Calcium	Ca	20	2,8,8,2

▼ **Table 9.3** Electronic configuration of the first three elements in Group VII

Element	Symbol	Proton number	Electronic configuration
Fluorine	F	9	2,7
Chlorine	Cl	17	2,8,7
Bromine	Br	35	2,8,18,7

The metallic character of the elements in a group increases as you move down the group. This is because electrons become easier to lose as the outer shell electrons become further from the nucleus. There is less attraction between the nucleus and the outer shell electrons because of the increased distance between them.

9.3 Group I – the alkali metals

Group I consists of the five metals lithium, sodium, potassium, rubidium and caesium, and the radioactive element francium. Lithium, sodium and potassium are commonly available for use in school. They are all very reactive metals and they are stored under oil to prevent them coming into contact with water or air. These three metals have the following properties.

» They are good conductors of electricity and heat.
» They are soft metals. Lithium is the hardest and potassium the softest.

» They are metals with low densities. For example, lithium has a density of 0.53 g/cm³ and potassium has a density of 0.86 g/cm³.

» They have shiny surfaces when freshly cut with a knife (Figure 9.6).

» They have low melting points. For example, lithium has a melting point of 181°C and potassium has a melting point of 64°C.

» They burn in oxygen or air, with characteristic flame colours, to form white solid oxides. For example, lithium reacts with oxygen in air to form white lithium oxide, according to the following equation:

lithium + oxygen → lithium oxide

$$4\text{Li(s)} + \text{O}_2\text{(g)} \rightarrow 2\text{Li}_2\text{O(s)}$$

▲ **Figure 9.6** Cutting sodium metal

a   Potassium reacts very vigorously with cold water

b   An alkaline solution is produced when potassium reacts with water

▲ **Figure 9.7**

These Group I oxides all react with water to form alkaline solutions of the metal hydroxide.

lithium oxide + water → lithium hydroxide

$$Li_2O(s) + H_2O(l) \rightarrow 2LiOH(aq)$$

» They react vigorously with water to give an alkaline solution of the metal hydroxide as well as producing hydrogen gas. For example:

potassium + water → potassium + hydrogen gas
hydroxide

$$2K(s) + 2H_2O(l) \rightarrow 2KOH(aq) + H_2(g)$$

» Of the first three metals in Group I, potassium is the most reactive towards water (Figure 9.7), followed by sodium and then lithium. Such gradual changes we call **trends**. Trends are useful to chemists as they allow predictions to be made about elements we have not observed in action.

» They react vigorously with halogens, such as chlorine, to form metal halides, for example sodium chloride (Figure 9.8).

sodium + chlorine → sodium chloride

$$2Na(s) + Cl_2(g) \rightarrow 2NaCl(s)$$

Considering the group as a whole, the further down the group you go, the more reactive the metals become. Francium is, therefore, the most reactive Group I metal.

Table 9.1 shows the electronic configuration of the first three elements of Group I. You will notice in each case that the outer shell contains only one electron. When these elements react they lose this outer electron, and in doing so, become more stable, because they obtain the electronic configuration of a noble gas. You will learn more about the stable nature of these gases later in this chapter.

When, for example, the element sodium reacts, it loses its outer electron. This requires energy to overcome the electrostatic attractive forces between the outer electron and the positive nucleus (Figure 9.9).

▲ **Figure 9.8** A very vigorous reaction takes place when sodium burns in chlorine gas. Sodium chloride is produced

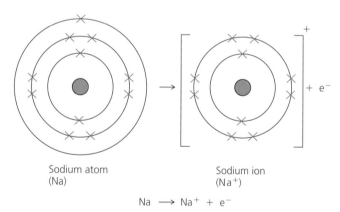

Sodium atom
(Na)

Sodium ion
(Na⁺)

$$Na \rightarrow Na^+ + e^-$$

▲ **Figure 9.9** This sodium atom loses an electron to become a sodium ion

Look at Figure 9.10. Why do you think potassium is more reactive than lithium or sodium?

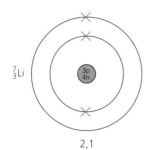

$^{7}_{3}Li$

2,1

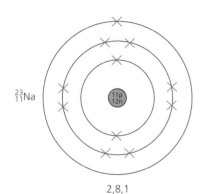

$^{23}_{11}Na$

2,8,1

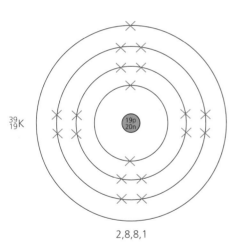

$^{39}_{19}K$

2,8,8,1

▲ **Figure 9.10** Electronic configuration of lithium, sodium and potassium

Potassium is more reactive because less energy is required to remove the outer electron from its atom than for lithium or sodium. This is because as you go down the group, the size of the atoms increases and the outer electron gets further away from the nucleus and becomes easier to remove.

Test yourself

1 Write word and balanced chemical equations for the reactions between:
   a sodium and oxygen
   b sodium and water.
2 a Using the information on p. 132–134, predict the properties of the element francium related to its melting point, density and softness.
   b Predict how francium would react with water and write a balanced equation for the reaction.
3 Write word and balanced chemical equations for the reactions between:
   a magnesium and water
   b calcium and oxygen.
4 Explain the fact that calcium is more reactive than magnesium in terms of their electronic configurations.

➔ Going further

Group II – the alkaline earth metals

Group II consists of the five metals beryllium, magnesium, calcium, strontium and barium, and the radioactive element radium. Magnesium and calcium are generally available for use in school. These metals have the following properties.

● They are harder than those in Group I.
● They are silvery-grey in colour when pure and clean. They tarnish quickly, however, when left in air due to the formation of a metal oxide on their surfaces (Figure 9.11).

▲ **Figure 9.11** Tarnished (left) and cleaned-up magnesium

● They are good conductors of heat and electricity.
● They burn in oxygen or air with characteristic flame colours to form solid white oxides. For example:

magnesium + oxygen → magnesium oxide

$$2Mg(s) + O_2(g) \rightarrow 2MgO(s)$$

- They react with water, but they do so much less vigorously than the elements in Group I. For example:

calcium + water → calcium hydroxide + hydrogen gas

$$Ca(s) + 2H_2O(l) \rightarrow Ca(OH)_2(aq) + H_2(g)$$

Considering the group as a whole, the further down the group you go, the more reactive the elements become.

9.4 Group VII – the halogens

Group VII consists of the four elements fluorine, chlorine, bromine and iodine, and the radioactive element astatine. Of these five elements, chlorine, bromine and iodine are generally available for use in school.

a   Chlorine, bromine and iodine

» These elements are coloured and become darker going down the group (Table 9.4).

▼ **Table 9.4** Colours of some halogens

Halogen	Colour
Chlorine	Pale yellow-green gas
Bromine	Red-brown liquid
Iodine	Grey-black solid

» They exist as diatomic molecules, for example Cl_2, Br_2 and I_2.
» At room temperature and pressure they show a gradual change from a gas (Cl_2), through a liquid (Br_2), to a solid (I_2) (Figure 9.12) as the density increases.
» They form molecular compounds with other non-metallic elements, for example HCl.
» They react with hydrogen to produce the hydrogen halides, which dissolve in water to form acidic solutions.

hydrogen + chlorine → hydrogen chloride

$$H_2(g) + Cl_2(g) \rightarrow 2HCl(g)$$

hydrogen + water → hydrochloric chloride   acid

$$HCl(g) + H_2O \rightarrow HCl(aq) \rightarrow H^+(aq) + Cl^-(aq)$$

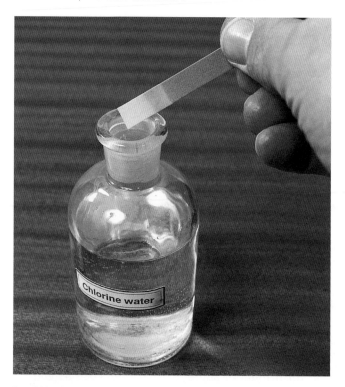

b   Chlorine gas bleaches moist indicator paper
▲ **Figure 9.12**

» They react with metals to produce ionic metal halides, for example chlorine and iron produce iron(III) chloride.

iron + chlorine → iron(III) chloride

$$2Fe(s) + 3Cl_2(g) \rightarrow 2FeCl_3(s)$$

Displacement reactions

If chlorine is bubbled into a solution of potassium iodide, the less reactive halogen, iodine, is displaced by the more reactive halogen, chlorine, as you can see from Figure 9.13:

potassium + chlorine → potassium + iodine
iodide   chloride

$$2KI(aq) + Cl_2(g) \rightarrow 2KCl(aq) + I_2(aq)$$

▲ **Figure 9.13** Iodine being displaced from potassium iodide solution as chlorine is bubbled through

The observed order of reactivity of the halogens, confirmed by similar **displacement reactions**, is:

Decreasing reactivity
chlorine bromine iodine →

You will notice that, in contrast to the elements of Groups I and II, the order of reactivity decreases on going down the group.

Table 9.5 shows the electronic configuration for chlorine and bromine. In each case the outer shell contains seven electrons. When these elements react, they gain one electron per atom to gain the stable electronic configuration of a noble gas. You will learn more about the stable nature of these gases in the next section. For example, when chlorine reacts it gains a single electron and forms a negative ion (Figure 9.14).

▼ **Table 9.5** Electronic configuration of chlorine and bromine

Element	Symbol	Proton number	Electronic configuration
Chlorine	Cl	17	2,8,7
Bromine	Br	35	2,8,18,7

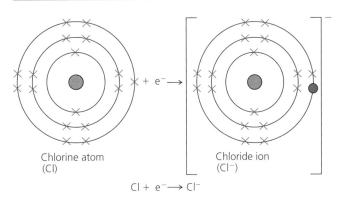

Chlorine atom (Cl)   Chloride ion (Cl⁻)

$$Cl + e^- \rightarrow Cl^-$$

▲ **Figure 9.14** A chlorine atom gains an electron to form a chloride ion

Chlorine is more reactive than bromine because the incoming electron is gained more easily by the smaller chlorine atom than in the larger bromine atom. It is gained more easily because there is a stronger attraction between the negative charge of the incoming electron and the positive charge of the nucleus. In the larger bromine atom, there are more occupied electron shells surrounding the nucleus, which lessen the attraction of the nucleus and the electrons in these shells repel the incoming electron. This makes it harder for the bromine atom to gain the extra electron it needs to gain a stable electronic configuration. This is the reason the reactivity of the halogens decreases going down the group.

The halogens and their compounds are used in many different ways (Figure 9.15).

In the reaction of chlorine with potassium iodide, both Cl atoms in Cl_2 gain an electron from an iodide ion, I^-, thus forming two chloride ions, Cl^-. The iodine atoms formed by the loss of an electron combine to give an iodine molecule, I_2.

Oxidation: $2I^-(aq) + Cl_2(g) \rightarrow 2Cl^-(aq) + I_2(aq)$

Number: −1   0   −1   0

Colour: colourless green   colourless orange-
                                                brown

The iodide ion has been oxidised because it has lost electrons. The oxidation number has increased. Chlorine has been reduced because it has gained electrons. The oxidation number has decreased.

Going further

Uses of the halogens

- Fluorine is used in the form of fluorides in drinking water and toothpaste because it reduces tooth decay by hardening the enamel on teeth.
- Chlorine is used to make PVC plastic as well as household bleaches. It is also used to kill bacteria and viruses in drinking water (Chapter 11, p. 168).
- Bromine is used to make disinfectants, medicines and fire retardants.
- Iodine is used in medicines and disinfectants, and also as a photographic chemical.

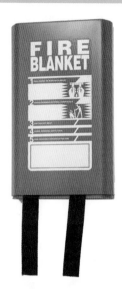

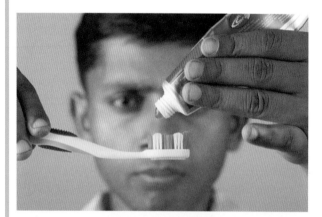

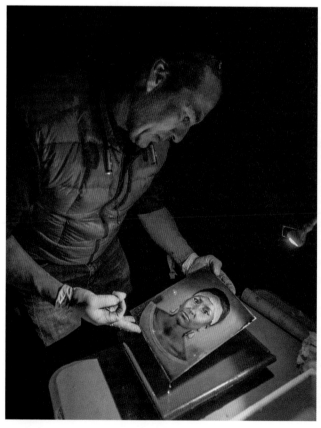

▲ **Figure 9.15** The halogens have many varied uses – fluoride in toothpaste to help reduce dental decay, iodine as a photographic chemical, chlorine in household bleach to kill bacteria, bromine as a fire retardant

Practical skills

Halogen displacement reactions

Safety

- Eye protection must be worn.
- Take care with halogen water solutions (harmful).
- Take care with organic solvents (may be harmful)

Displacement reactions can be used to determine the reactivity of the Group VII elements, the halogens. In this experiment, a student uses these reactions to determine an order of reactivity for iodine, chlorine and bromine, and predicts the reactivity of the other two halogens, fluorine and astatine.

If a more reactive halogen reacts with a compound of a less reactive halogen, the less reactive halogen will be displaced and it will form the halogen molecule, while the more reactive halogen becomes a halide ion.

To observe the presence of the different halogens in solution, the student used an organic solvent as the halogens produce more vivid colours in this solvent compared with water.

To show the colour of the halogens in the organic solvent solutions, the student separately placed three halogens in test tubes and added a small amount of the organic solvent. The tubes were then fitted with a rubber bung and shaken.

After shaking, two layers formed (Figure 9.16). The lower layer is the water (aqueous layer) and the upper layer is the organic layer.

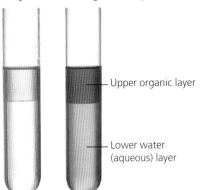

▲ **Figure 9.16** The tube on the left shows the results of chlorine water being added to sodium bromide, and on the right the result of bromine water being added to sodium iodide

The results obtained are:

Halogen	Colour in water	Colour in organic solvent
Chlorine	Colourless	Pale green
Bromine	Orange	Orange
Iodine	Brown	Violet

The colours of the halogens in water can be used to identify if a reaction has occurred when a solution of a halogen, for example, chlorine water, is mixed with a solution of a sodium halide, for example, sodium iodide solution. The table below shows the results of six of these.

Sample data

		Chlorine water	Bromine water	Iodine water
Colour after shaking with sodium iodide solution		Brown	Brown	
Colour of each layer with the organic solvent added	Upper	Violet	Violet	
	Lower	Brown	Brown	
Colour after shaking with sodium bromide solution		Orange		Brown
Colour of each layer with the organic solvent added	Upper	Orange		Violet
	Lower	Pale orange		Brown
Colour after shaking with sodium chloride solution			Orange	Brown
Colour of each layer with the organic solvent added	Upper		Orange	Violet
	Lower		Pale orange	Brown

1 In how many of the six reactions has a displacement reaction occurred?
2 For each of these displacement reactions, write a balanced chemical equation.
3 What type of chemical change has happened to the halogen molecules in the displacement reactions?
4 State the order of reactivity of the three halogens used in the experiment.
5 What would be the colour of fluorine and astatine in water?
6 Which of the five halogen elements is the most reactive? Explain your answer.
7 Why were solutions of the halogens used, rather than the halogens themselves?

9.5 Group VIII – the noble gases

Helium, neon, argon, krypton, xenon and the radioactive element radon make up a most unusual group of non-metals, called the noble gases. They were all discovered after Mendeleev had published his periodic table. They were discovered between 1894 and 1900, mainly through the work of the British scientists Sir William Ramsay and Lord John William Strutt Rayleigh.

» They are colourless gases.
» They are monatomic gases – they exist as individual atoms, for example He, Ne and Ar.
» They are very unreactive.

No compounds of helium, neon or argon have ever been found. However, more recently a number of compounds of xenon and krypton with fluorine and oxygen have been produced, for example XeF_6.

These gases are chemically unreactive because they have electronic configurations which are stable and very difficult to change (Table 9.6). They are so stable that other elements attempt to attain these electronic configurations during chemical reactions (Chapter 3, pp. 31 and 38). You have probably seen this in your study of the elements of Groups I, II and VII.

▼ **Table 9.6** Electronic configuration of helium, neon and argon

Element	Symbol	Proton number	Electronic configuration
Helium	He	2	2
Neon	Ne	10	2,8
Argon	Ar	18	2,8,8

Although unreactive, they have many uses. Argon, for example, is the gas used to fill light bulbs to prevent the tungsten filament reacting with air. Neon is used extensively in advertising signs and in lasers.

> **Test yourself**
>
> 5 Write word and balanced chemical equations for the reactions between:
> a bromine and potassium iodide solution
> b bromine and potassium chloride solution.
> If no reaction takes place, write 'no reaction' and explain why.
> c Using the information on pp. 136–137, predict the properties of the element astatine related to its melting point, density and physical state at room temperature. Predict how astatine would react with sodium bromide solution.

9.6 Transition elements

This block of metals includes many you will be familiar with, for example copper, iron, nickel, zinc and chromium (Figure 9.17).

» They are less reactive metals.
» They form a range of brightly coloured compounds (Figure 9.18).
» They are harder and stronger than the metals in Groups I and II.
» They have much higher densities than the metals in Groups I and II.
» They have high melting points (except for mercury, which is a liquid at room temperature).
» They are good conductors of heat and electricity.
» They show catalytic activity (Chapter 7) as elements and compounds. For example, iron is used in the industrial production of ammonia gas (see Haber process, Chapter 7, p. 102).
» They do not react (corrode) so quickly with oxygen and/or water.

» They form simple ions with variable oxidation numbers. (For a discussion of oxidation number see Chapter 3.)
For example:
- Copper forms Cu^+ (Cu(I)) and Cu^{2+} (Cu(II)), in compounds such as Cu_2O and $CuSO_4$.
- Iron forms Fe^{2+} (Fe(II)) and Fe^{3+} (Fe(III)), in compounds such as $FeSO_4$ and $FeCl_3$.
- Cobalt forms Co^{2+} (Co(II)) and Co^{3+} (Co(III)), in compounds such as $CoCl_2$ and $Co(OH)_3$.

» They form more complicated ions with high oxidation numbers. For ezxample, chromium forms the dichromate(VI) ion, $Cr_2O_7^{2-}$, which contains chromium with a +6 oxidation number (Cr(VI)) and manganese forms the manganate(VII) ion, MnO_4^-, which contains manganese with a +7 oxidation number (Mn(VII)) (see Chapter 3).

a Copper is used in many situations which involve good heat and electrical conduction. It is also used in medallions and bracelets

b These gates are made of iron. Iron can easily be moulded into different shapes

d This bucket has been coated with zinc to prevent the steel of the bucket corroding

c Monel is an alloy of nickel and copper. It is extremely resistant to corrosion, even that caused by sea water

e The alloy stainless steel contains a high proportion of chromium, which makes it corrosion resistant

▲ **Figure 9.17** Everyday uses of transition elements and their compounds. They are often known as the 'everyday metals'

a Some solutions of coloured transition element compounds

b The coloured compounds of transition elements can be seen in these pottery glazes

▲ **Figure 9.18**

> **Test yourself**
>
> 6 Look at the photographs in Figure 9.17 and state which key properties are important when considering the particular use of the metal.
> 7 Which groups in the Periodic Table contain:
>   **a** only metals?
>   **b** only non-metals?
>   **c** both metals and non-metals?
> 8 Give the oxidation number of the transition metal ion in:
>   **a** FeO
>   **b** Fe_2O_3
>   **c** CuO
>   **d** $Co(OH)_2$

9.7 The position of hydrogen

Hydrogen is often placed by itself in the Periodic Table. This is because the properties of hydrogen are unique. However, useful comparisons can be made with the other elements. It is often shown at the top of either Group I or Group VII, but it cannot fit easily into the trends shown by either group; see Table 9.7.

▼ **Table 9.7** Comparison of hydrogen with lithium and fluorine

Lithium	Hydrogen	Fluorine
Solid	Gas	Gas
Forms a positive ion	Forms positive or negative ions	Forms a negative ion
1 electron in outer shell	1 electron in outer shell	1 electron short of a full outer shell
Loses 1 electron to form a noble gas configuration	Needs 1 electron to form a noble gas configuration	Needs 1 electron to form a noble gas configuration

Revision checklist

After studying Chapter 9 you should be able to:
✓ Describe the arrangement of elements in groups and periods in the Periodic Table.
✓ Use the Periodic Table to help you to predict the properties of elements.
✓ Use the Periodic Table to help you to give the charges on ions and to write chemical and ionic equations.
✓ Describe the trends in the properties and reactions of the Group I and Group VII elements.
✓ Describe the Group VIII elements as unreactive, monoatomic gases and explain why they are unreactive.
✓ Identify trends in the different groups if you are given information about physical or chemical properties.
✓ Describe the physical appearance of the Group I and VII elements.
✓ Describe halogen-halide displacement reactions.
✓ Give the properties of the transition metals.
✓ Understand that transition metal ions have variable oxidation numbers.
✓ Describe the Group VIII elements as unreactive, monoatomic gases and explain why they are unreactive.

Exam-style questions

1 The diagram below shows part of the Periodic Table.

I	II												III	IV	V	VI	VII	0
		H																He
Li	Be												B			O		
														Si	P		Cl	Ar
	Ca												Zn	Ga			Se	Br

Using **only** the symbols of the elements shown above, give the symbol for an element which:

a is a pale yellow-green coloured toxic gas [1]
b is stored under oil [1]
c has five electrons in its outer electron energy shell [1]
d is the most reactive Group II element [1]
e is the most reactive halogen [1]
f is the only liquid shown [1]
g is a transition element [1]
h is a gas with two electrons in its outer shell. [1]

2 Three members of the halogens are: $^{35.5}_{17}Cl$, $^{80}_{35}Br$ and $^{127}_{53}I$.
a i Give the electronic configuration of an atom of chlorine. [1]
ii Explain why the relative atomic mass of chlorine is not a whole number. [1]
iii Identify how many protons there are in an atom of bromine. [1]
iv Identify how many neutrons there are in an atom of iodine. [1]
v Explain the order of reactivity of these elements. [5]
b When potassium is allowed to burn in a gas jar of chlorine, in a fume cupboard, clouds of white smoke are produced.
i Explain why this reaction is carried out in a fume cupboard. [1]
ii State what the white smoke consists of. [1]
iii Give a word and balanced chemical equation for this reaction. [2]
iv Describe what you would expect to see when potassium is allowed to burn safely in a gas jar of bromine vapour. Write a word and balanced chemical equation for this reaction. [3]

3 'By using displacement reactions it is possible to deduce the order of reactivity of the halogens'. Discuss this statement with reference to the elements bromine, iodine and chlorine only. [4]

4 Use the information given in the table below to answer the questions concerning the elements **Q**, **R**, **S**, **T** and **X**.

Element	Proton number	Mass number	Electronic configuration
Q	3	7	2,1
R	20	40	2,8,8,2
S	18	40	2,8,8
T	8	18	2,6
X	19	39	2,8,8,1

a Identify the element that has 22 neutrons in each atom. [1]
b Identify the element that is a noble gas. [1]
c Identify the two elements that form ions with the same electronic configuration as neon. [2]
d Identify the two elements that are in the same group of the Periodic Table and group this is. [2]
e Place the elements in the table into the periods in which they belong. [3]
f Identify the most reactive metal element in the table. [1]
g Identify which of the elements is calcium. [1]

5 a Consider the chemical properties and physical properties of the halogens: chlorine, bromine and iodine. Using these properties, predict the following about the other two halogens, fluorine and astatine. [2]

Property	Fluorine	Astatine
State at room temperature and pressure		
Colour		
Reactivity with sodium metal		

b i Give a word equation for the reaction of chlorine gas with sodium bromide solution. [1]

ii Give a balanced chemical equation for the reaction, with state symbols. [2]

iii Give an ionic equation for the reaction, with state symbols. [2]

6 Some of the most important metals we use are found in the transition element section of the Periodic Table. One of these elements is copper. Sodium, a Group I metal, has very different properties from those of copper. Complete the table to show their differences. [4]

	Transition element, e.g. copper	Group I metal, e.g. sodium
Hardness (hard/soft)		
Reactivity		
Density (high/low)		
Variable oxidation states		

10 Metals

FOCUS POINTS

★ What are the general physical and chemical properties of metals?
★ How do their physical properties affect their uses?
★ What is an alloy and how does it differ from a pure metal?
★ In what ways does the reactivity of metals affect their reactions and extraction?

In this chapter you will look at the extraction, properties and chemical reactions of metals. Some of these metallic elements are the most important elements in the world. They are used in the building of roads, bridges, factories, in the movement of electricity and as catalysts. With all these metals available to use, what properties make some better than others for a particular job?

We will look at the way in which we are able to modify the properties of metals to make them have different and better properties and how we try to protect metallic objects from corrosion to extend their useful life. Why is this important?

By the end of this chapter you should be able to predict the chemical properties of a metal to decide which metal would be best for a particular job.

10.1 Properties of metals

The majority of the elements in the Periodic Table are metals. They have very different properties to the non-metallic elements. A comparison of their properties is shown in Table 10.1.

▼ **Table 10.1** How the properties of metals and non-metals compare

Property	Metal	Non-metal
Thermal conductivity	Good	Poor
Electrical conductivity	Good	Poor
Malleability	Good	Poor – usually soft or brittle
Ductility	Good	Poor – usually soft or brittle
Melting point	Usually high	Usually low
Boiling point	Usually high	Usually low

a Sodium burning in air/oxygen

b Iron rusts when left unprotected

You have already seen in Chapter 2, p. 11, that metals usually have similar physical properties. However, they differ in other ways. Look closely at the three photographs in Figure 10.1.

c   Gold is used in leaf form on this giant Buddha as it is unreactive

▲ **Figure 10.1**

Sodium is soft and reacts violently with both air and water. Iron also reacts with air and water but much more slowly, forming **rust**. Gold, however, remains totally unchanged after many hundreds of years. Sodium is said to be more reactive than iron and, in turn, iron is said to be more reactive than gold.

10.2 Metal reactions

By carrying out reactions in the laboratory with other metals and with air, water and dilute acid, it is possible to produce an order of reactivity of the metals.

With acid

If a metal reacts with dilute hydrochloric acid then hydrogen and the metal chloride are produced. Look closely at the photograph in Figure 10.2 showing magnesium metal reacting with dilute hydrochloric acid. You will notice effervescence, which is caused by bubbles of hydrogen gas being formed as the reaction between the two substances proceeds. The other product of this reaction is the salt, magnesium chloride.

magnesium + hydrochloric → magnesium + hydrogen
acid   chloride

$$Mg(s) \quad + \quad 2HCl(aq) \quad \rightarrow \quad MgCl_2(aq) + \quad H_2(g)$$

If similar reactions are carried out using other metals with acid, an order of reactivity can be produced by measuring the rate of evolution of hydrogen. This is known as a **reactivity series**.

▲ **Figure 10.2** Effervescence occurs when magnesium is put into acid

> **Key definition**
>
> The **reactivity series** is the order of the reactivity of the following elements: potassium, sodium, calcium, magnesium, aluminium, carbon, zinc, iron, hydrogen, copper, silver, gold.

With air/oxygen

Many metals react directly with oxygen to form oxides. For example, magnesium burns brightly in oxygen to form the white powder magnesium oxide.

magnesium + oxygen → magnesium oxide

$$2Mg(s) \quad + \quad O_2(g) \quad \rightarrow \quad 2MgO(s)$$

With cold water/steam

Reactive metals such as potassium, sodium and calcium react with cold water to produce the metal hydroxide and hydrogen gas. For example, the reaction of sodium with water produces sodium hydroxide and hydrogen.

sodium + water → sodium hydroxide + hydrogen

$$2Na(s) + 2H_2O(l) \rightarrow \quad 2NaOH(aq) \quad + \quad H_2(g)$$

The moderately reactive metals, magnesium, zinc and iron, react slowly with water. They will, however, react more rapidly with steam (Figure 10.3). In their reaction with steam, the metal oxide and hydrogen are formed. For example, magnesium produces magnesium oxide and hydrogen gas.

magnesium + steam → magnesium oxide + hydrogen

$$Mg(s) + H_2O(g) → MgO(s) + H_2(g)$$

This experiment should only ever be carried out as a teacher demonstration. Eye protection must be worn during the demonstration. There is the danger of suckback of water into the boiling tube, so your teacher will remove the delivery tube from the water before heating is stopped.

An order of reactivity, giving the most reactive metal first, using results from experiments with dilute acid, is shown in Table 10.2. The table also shows how the metals react with air/oxygen and water/steam, and, in addition, the ease of extraction of the metal.

In all these reactions the most reactive metal is the one that has the highest tendency to lose outer electrons to form a positive metal ion.

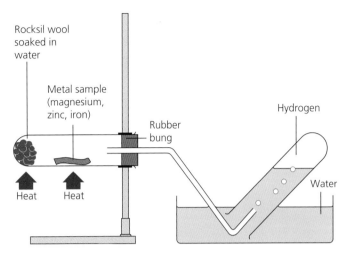

▲ **Figure 10.3** Apparatus used to investigate how metals such as magnesium react with steam

▼ **Table 10.2** Order of reactivity

Reactivity series	Reaction with dilute acid	Reaction with air/oxygen	Reaction with water	Ease of extraction	
Potassium (K) Sodium (Na)	Produce H₂ with decreasing vigour	Burn very brightly and vigorously	Produce H₂ with decreasing vigour with cold water	Difficult to extract	
Calcium (Ca) Magnesium (Mg)		Burn to form an oxide with decreasing vigour	React with steam with decreasing vigour	Easier to extract	Increasing reactivity of metal
Aluminium (Al*)					
[Carbon (C)]					
Zinc (Zn)					
Iron (Fe)					
[Hydrogen (H)]		React slowly to form the oxide	Do not react with cold water or steam		
Copper (Cu)	Do not react with dilute acids				
Silver (Ag)		Do not react		Found as the element (uncombined)	
Gold (Au)					

* Because aluminium reacts so readily with the oxygen in the air, a protective oxide layer is formed on its surface. This often prevents any further reaction and disguises aluminium's true reactivity. This gives us the use of a light and strong metal.

a This wood-burning stove is made of iron

b Copper pots and pans

▲ **Figure 10.4**

▲ **Figure 10.5** Planes are made of an alloy which contains magnesium and aluminium

Generally, it is the unreactive metals that we find the most uses for; for example, the metals iron and copper can be found in many everyday objects (Figure 10.4). However, magnesium is one of the metals used in the construction of the Airbus A380 (Figure 10.5).

Aluminium, although a reactive metal, has a protective oxide coating on its surface which forms when it is manufactured. This allows it to be used in many areas of industry, such as in the manufacture of cars, aircraft and overhead electricity cables, because of its low density and good electrical conductivity.

Both sodium and potassium are so reactive that they have to be stored under oil to prevent them from coming into contact with water or air. However, because they have low melting points and are good conductors of heat, they are used as coolants for nuclear reactors.

> **Test yourself**
>
> 1 Write balanced chemical equations for the reactions between:
>   a iron and dilute hydrochloric acid
>   b potassium and water
>   c zinc and oxygen
>   d calcium and water.
> 2 Make a list of six things you have in your house made from copper or iron.
>   Give a use for each of the other unreactive metals shown in the reactivity series.

10.3 Reactivity of metals and their uses

Generally, it is the unreactive metals for which we find the most uses. For example, copper is used for electrical wiring because of its good electrical conductivity and ductility. However, the metal aluminium is an exception. Aluminium appears in the reactivity series just below magnesium and is quite reactive.

Fortunately, it forms a relatively thick oxide layer on the surface of the metal which prevents further reaction. This gives us a light, strong metal for use in:

>> the manufacture of aircraft because of its low density
>> the manufacture of overhead electrical cables because of its low density and good electrical conductivity
>> food containers because of its resistance to corrosion.

> **Test yourself**
>
> 3 Compare the uses of magnesium, a reactive metal, and lead, an unreactive metal in terms of their reactivity and properties.

Displacement reactions

Metals compete with each other for other anions, in solution. This type of reaction is known as a displacement reaction. As in the previous type of competitive reaction, the reactivity series can be used to predict which of the metals will 'win'.

In a displacement reaction, a more reactive metal will displace a less reactive metal from a solution of its salt. Zinc is above copper in the reactivity series. Figure 10.6 shows what happens when a piece of zinc metal is left to stand in a solution of copper(II) nitrate. The copper(II) nitrate slowly loses its blue colour as the zinc continues to displace the copper from the solution and eventually becomes colourless zinc nitrate.

$$\text{zinc} + \text{copper(II) nitrate} \rightarrow \text{zinc nitrate} + \text{copper}$$

$$Zn(s) + Cu(NO_3)_2(aq) \rightarrow Zn(NO_3)_2(aq) + Cu(s)$$

The ionic equation for this reaction is:

$$\text{zinc} + \text{copper ions} \rightarrow \text{zinc ions} + \text{copper}$$

$$Zn(s) + Cu^{2+}(aq) \rightarrow Zn^{2+}(aq) + Cu(s)$$

This is also a redox reaction involving the transfer of two electrons from the zinc metal to the copper ions. The zinc is oxidised to zinc ions in aqueous solution, while the copper ions are reduced. (See Chapter 5, p. 68, for a discussion of oxidation and reduction in terms of electron transfer.) It is possible to confirm the reactivity series for metals using reactions of the types discussed in this section.

▲ **Figure 10.6** Zinc displaces copper

Practical skills

Using displacement reactions to determine the order of reactivity

Safety

- Eye protection must be worn.
- Metal nitrate solutions (≤ 1M) may be harmful.

In a displacement reaction a more reactive metal will displace a less reactive metal from a solution of its salt. It is possible to carry out displacement reactions using a dimple tile (see Figure 10.7).

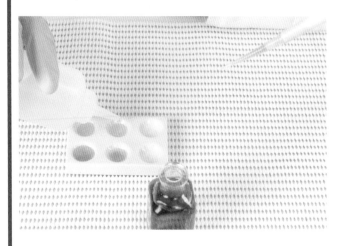

▲ **Figure 10.7** Dimple tile

Small pieces of metals can be placed into the dimples and then, using a dropping pipette, a solution containing another metal ion can be put over the metal. It is usual to use metal nitrate solutions as they are all soluble.

After the metal solution has been added to the metal you must ensure that the metal is below the surface of the solution and leave time for the reaction to occur: usually 2 or 3 minutes is adequate.

1 Why is it important that the metal is under the surface of the solution?
2 What would be observed if a reaction has occurred?

Sample data

Five metals and nitrate solutions of those metals were reacted together using the method above. The five metals were tin, magnesium, zinc, copper and iron. The results obtained were collected in the table below. A tick (✓) indicates a colour change was observed, a cross (x) indicates no colour change was observed.

3 Using the information in the table, state which of the metals was the most reactive.
4 Using the information in the table, state which of the metals was the least reactive.
5 Arrange the five metals in order of increasing reactivity.
6 a Write a word, balanced chemical and ionic equation for the reaction between magnesium metal and copper(II) nitrate.
  b What would you observe when this reaction occurs?

	Nitrates				
	Copper	**Zinc**	**Tin**	**Magnesium**	**Iron**
Copper	–	x	x	x	x
Zinc	✓	–	✓	x	✓
Tin	✓	x	–	x	x
Magnesium	✓	✓	✓	–	✓
Iron	✓	x	✓	x	–

> **Test yourself**
>
> 4 Predict whether or not the following reactions will take place:
>   a magnesium + calcium nitrate solution
>   b iron + copper(II) nitrate solution
>   c copper + silver nitrate solution.
>   Complete the word equations and write balanced chemical and ionic equations for those reactions which do take place.
> 5 How could you carry out a series of reactions between metals and solutions of their nitrates to establish a reactivity series?
> 6 The data below was obtained by carrying out displacement reactions of five metals with the nitrates of the same five metals. Strips of each metal were placed in solutions of the other four metals nitrate solutions.
>
> | | Nitrates | | | | |
> |---|---|---|---|---|---|
> | | **A** | **B** | **C** | **D** | **E** |
> | **A** | — | ✓ | ✓ | ✓ | |
> | **B** | x | — | x | x | x |
> | **C** | x | ✓ | — | ✓ | x |
> | **D** | x | ✓ | x | — | x |
> | **E** | ✓ | ✓ | ✓ | ✓ | — |
>
> ✓ = metal displaced   x = no reaction
> Put the five metals **A–E** in order of their reactivity using the data above.

10.4 Identifying metal ions

When an alkali dissolves in water, it produces hydroxide ions. It is known that most metal hydroxides are insoluble. So if hydroxide ions from a solution of an alkali are added to a solution of a metal salt, an insoluble, often coloured, metal hydroxide is precipitated from solution (Figure 10.8).

If we take the example of iron(III) chloride with sodium hydroxide solution:

$$\begin{array}{ccc} \text{chloride} + \text{hydroxide} & \rightarrow & \text{hydroxide} + \text{chloride} \\ \text{iron(III)} \quad \text{sodium} & & \text{sodium} \quad \text{iron(III)} \end{array}$$

$$FeCl_3(aq) + 3NaOH(aq) \rightarrow Fe(OH)_3(s) + 3NaCl(aq)$$

a Iron(III) hydroxide is precipitated

b Copper(II) hydroxide is precipitated

▲ **Figure 10.8**

The ionic equation for this reaction is:

iron(III) ions + hydroxide ions → iron(III) hydroxide

$$Fe^{3+}(aq) \quad + \quad 3OH^-(aq) \quad \rightarrow \quad Fe(OH)_3(s)$$

Table 10.3 shows the effects of adding a few drops of sodium hydroxide solution to solutions containing various metal ions, and of adding an excess. The colours of the insoluble metal hydroxides can be used to identify the metal cations present in solution. In some cases, the precipitate dissolves in excess hydroxide, owing to the amphoteric nature of the metal hydroxide. This amphoteric nature can also be used to help identify metals such as aluminium and zinc.

▼ **Table 10.3** The effect of adding sodium hydroxide solution to solutions containing various metal ions

Metal ion present in solution	Effect of adding sodium hydroxide solution	
	A few drops	**An excess**
Aluminium, Al^{3+}	White precipitate of aluminium hydroxide	Precipitate is soluble in excess, giving a colourless solution
Calcium, Ca^{2+}	White precipitate of calcium hydroxide	Precipitate is insoluble in excess
Chromium(III), Cr^{3+}	Green precipitate of chromium(III) hydroxide	Precipitate is insoluble in excess
Copper(II), Cu^{2+}	Light blue precipitate of copper(II) hydroxide	Precipitate is insoluble in excess
Iron(II), Fe^{2+}	Green precipitate of iron(II) hydroxide	Precipitate is insoluble in excess, turns brown near the surface on standing
Iron(III), Fe^{3+}	Red-brown precipitate of iron(III) hydroxide	Precipitate is insoluble in excess
Zinc, Zn^{2+}	White precipitate of zinc hydroxide	Precipitate is soluble in excess, giving a colourless solution

Test yourself

7 Write ionic equations for the reactions which take place to produce the metal hydroxides shown in Table 10.3.
8 Describe what you would see when sodium hydroxide is added slowly to a solution containing iron(II) nitrate.

10.5 Extraction of metals

Metals have been used since prehistoric times. Many primitive iron tools have been excavated. These were probably made from small amounts of native iron found in rock from meteorites. It was not until about 2500 BC that iron became more widely used. This date marks the dawn of the Iron Age, when people learned how to get iron from its ores in larger quantities by reduction using charcoal. An ore is a naturally occurring mineral from which a metal can be extracted. Metals lower in the reactivity series (such as iron and copper) tend to be easier to extract from their ores than more reactive metals (such as potassium and sodium).

The majority of metals are too reactive to exist on their own in the Earth's crust, and they occur naturally in rocks as compounds in ores (Figure 10.9). These ores are usually carbonates, oxides or sulfides of the metal, mixed with impurities.

Some metals, such as gold and silver, occur in a native form as the free metal (Figure 10.10). They are very unreactive and have withstood the action of water and the atmosphere for many thousands of years without reacting to become compounds.

Some of the common ores are shown in Table 10.4.

▲ **Figure 10.10** Gold crystals

Large lumps of the ore are first crushed and ground up by very heavy machinery. Some ores are already fairly concentrated when mined. For example, in some parts of the world, hematite contains over 80% Fe_2O_3. However, other ores, such as copper pyrites, are often found to be less concentrated, with only 1% or less of the copper compound, and so they have to be concentrated before the metal can be extracted. The method used to extract the metal from its ore depends on the position of the metal in the reactivity series.

▲ **Figure 10.9** Chalcopyrite, an ore of copper (top) and galena, an ore of lead (bottom)

▼ **Table 10.4** Some common ores

Metal	Name of ore	Chemical name of compound in ore	Formula	Usual method of extraction
Aluminium	Bauxite	Aluminium oxide	$Al_2O_3.2H_2O$	Electrolysis of oxide dissolved in molten cryolite
Copper	Copper pyrites	Copper iron sulfide	$CuFeS_2$	The sulfide ore is roasted in air
Iron	Hematite	Iron(III) oxide	Fe_2O_3	Heat oxide with carbon
Sodium	Rock salt	Sodium chloride	NaCl	Electrolysis of molten sodium chloride
Zinc	Zinc blende	Zinc sulfide	ZnS	Sulfide is roasted in air and the oxide produced is heated with carbon

Because reactive metals, such as sodium, hold on to the element(s) they have combined with, they are usually difficult to extract. For example, sodium chloride (as rock salt) is an ionic compound with the Na^+ and Cl^- ions strongly bonded to one another. The separation of these ions and the subsequent isolation of the sodium metal is therefore difficult.

Electrolysis of the molten, purified ore is the method used in these cases. During this process, the metal is produced at the cathode while a non-metal is produced at the anode. As you might expect, extraction of metal by electrolysis is expensive. In order to keep costs low, many metal smelters using electrolysis are situated in regions where there is hydroelectric power.

For further discussion of the extraction of aluminium, see Chapter 5, pp. 69–71.

Metals towards the middle of the reactivity series, such as iron and zinc, may be extracted by reducing the metal oxide with the non-metal carbon.

Extraction of iron

Iron is extracted mainly from its oxides, hematite (Fe_2O_3) and magnetite (Fe_3O_4), in a blast furnace (Figures 10.11 and 10.12). These ores contain at least 60% iron. The blast furnace is a steel tower, approximately 50 m high, lined with heat-resistant bricks. It is loaded with the 'charge' of iron ore (usually hematite), carbon in the form of coke (made by heating coal) and limestone (calcium carbonate).

A blast of hot air is sent in near the bottom of the furnace through holes which makes the 'charge' glow, as the coke burns in the preheated air.

carbon + oxygen → carbon dioxide

$$C(s) + O_2(g) \rightarrow CO_2(g)$$

▲ **Figure 10.11** A blast furnace

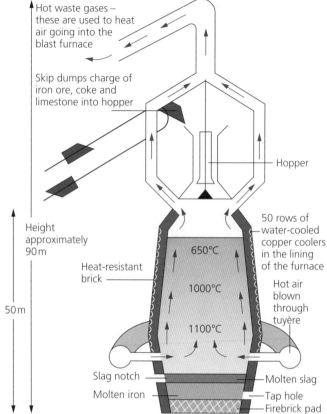

Hot waste gases – these are used to heat air going into the blast furnace

Skip dumps charge of iron ore, coke and limestone into hopper

Hopper

50 rows of water-cooled copper coolers in the lining of the furnace

Height approximately 90 m

Heat-resistant brick

650°C

1000°C

Hot air blown through tuyère

50 m

1100°C

Slag notch

Molten iron

Molten slag

Tap hole

Firebrick pad

▲ **Figure 10.12** Cross-section of a blast furnace

A number of chemical reactions then follow.

» The limestone begins to decompose:

calcium → calcium + carbon
carbonate   oxide   dioxide

$$CaCO_3(s) \rightarrow CaO(s) + CO_2(g)$$

» The carbon dioxide gas produced reacts with more hot coke higher up in the furnace, producing carbon monoxide in an endothermic reaction.

carbon dioxide + coke → carbon monoxide

$$CO_2(g) + C(s) \rightarrow 2CO(g)$$

» Carbon monoxide is a reducing agent (Chapter 3, p. 36). It rises up the furnace and reduces the iron(III) oxide ore. This takes place at a temperature of around 700°C:

iron(III) + carbon → iron + carbon
oxide   monoxide   dioxide

$$Fe_2O_3(s) + 3CO(g) \rightarrow 2Fe(s) + 3CO_2(g)$$

The molten iron trickles to the bottom.
» Calcium oxide is a base and this reacts with acidic impurities such as silicon(IV) oxide in the iron, to form **slag** which is mainly calcium silicate.

calcium oxide + silicon(IV) oxide → calcium silicate

$$CaO(s) + SiO_2(s) \rightarrow CaSiO_3(s)$$

The slag trickles to the bottom of the furnace, but because it is less dense than the molten iron, it floats on top of it.

Generally, metallic oxides, such as calcium oxide (CaO), are basic whereas non-metallic oxides, such as silicon(IV) oxide (SiO_2), are acidic.

Certain oxides, such as carbon monoxide (CO), are neutral and others, such as zinc oxide (ZnO), are amphoteric.

The molten iron, as well as the molten slag, may be tapped off (run off) at regular intervals.

The waste gases, mainly nitrogen and oxides of carbon, escape from the top of the furnace. They are used in a heat exchange process to heat incoming air and so help to reduce the energy costs of the process. Slag is the other waste material. It is used by builders and road makers (Figure 10.13) for foundations.

▲ **Figure 10.13** Slag is used in road foundations

The extraction of iron is a continuous process and is much cheaper to run than an electrolytic method.

▲ **Figure 10.14** The pouring of molten steel

Test yourself

9 How does the method used for extracting a metal from its ore depend on the metal's position in the reactivity series?

10 'It is true to say that almost all the reactions by which a metal is extracted from its ore are reduction reactions'. Discuss this statement with respect to the extraction of iron and aluminium.

Going further

Metal waste

Recycling has become commonplace in recent years (Figure 10.15). Why should we really want to recycle metals? Certainly, if we extract fewer metals from the Earth then the existing reserves will last that much longer. Also, recycling metals prevents the creation of a huge environmental problem (Figure 10.16). However, one of the main considerations is that it saves money.

▲ **Figure 10.15** Aluminium can recycling

The main metals which are recycled include aluminium and iron. Aluminium is saved by many households as drinks cans and milk bottle tops, to be melted down and recast. Iron is collected at community tips in the form of discarded household goods and it also forms a large part of the materials collected by scrap metal dealers. Iron is recycled to steel. Many steel-making furnaces run mainly on scrap iron.

Aluminium is especially easy to recycle at low cost. Recycling uses only 5% of the energy needed to extract the metal by electrolysis from bauxite. More than one-third of the world's need for aluminium is obtained by recycling.

▲ **Figure 10.16** If we did not recycle metals, then this sight would be commonplace

10.6 Metal corrosion

After a period of time, objects made of iron or steel will become coated with rust. The **rusting** of iron is a serious problem and wastes enormous amounts of money each year. Estimates are difficult to make, but it is thought that more than $2.5 trillion a year is spent worldwide on replacing iron and steel structures.

Rust is an orange–red powder consisting mainly of hydrated iron(III) oxide ($Fe_2O_3.xH_2O$). Both water and oxygen are essential for iron to rust, and if one of these two substances is not present then rusting will not take place. The rusting of iron is encouraged by salt. Figure 10.17 (p. 157) shows an experiment to show that oxygen (from the air) and water are needed for iron to rust.

> **Key definition**
> **Rusting** of iron and steel to form hydrated iron(II) oxide requires both water and oxygen.

Rust prevention

To prevent iron rusting, it is necessary to stop oxygen (from the air) and water coming into contact with it. There are several ways of doing this.

Painting

Ships, lorries, cars, bridges and many other iron and steel structures are painted to prevent rusting (Figure 10.18). However, if the paint is scratched, the iron beneath it will start to rust (Figure 10.19) and corrosion can then spread under the paintwork which is still sound. This is why it is essential that the paint is kept in good condition and checked regularly.

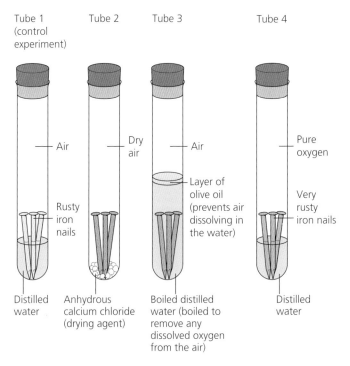

Tube 1 (control experiment) — Air — Rusty iron nails — Distilled water

Tube 2 — Dry air — Anhydrous calcium chloride (drying agent)

Tube 3 — Air — Layer of olive oil (prevents air dissolving in the water) — Boiled distilled water (boiled to remove any dissolved oxygen from the air)

Tube 4 — Pure oxygen — Very rusty iron nails — Distilled water

▲ **Figure 10.17** Rusting experiment with nails

▲ **Figure 10.18** Painting keeps the air and water away from the steel used to build a ship

▲ **Figure 10.19** A brand new car is protected against corrosion (top). However, if the paintwork is damaged, then rusting will result

Oiling/greasing

The iron and steel in the moving parts of machinery are coated with oil to prevent them from coming into contact with air or moisture. This is the most common way of protecting moving parts of machinery, but the protective film must be renewed.

Coating with plastic

The exteriors of refrigerators, freezers and many other items are coated with plastic, such as PVC, to prevent the steel structure rusting (Figure 10.20).

▲ **Figure 10.20** A coating of plastic stops metal objects coming into contact with oxygen or water

Galvanising

Some steel beams, used in the construction of bridges and buildings, are galvanised; steel waste collection bins are also galvanised. This involves dipping the object into molten zinc. The thin layer of the more reactive zinc metal coating the steel object slowly corrodes and loses electrons to the iron, thereby protecting it. This process continues even when much of the layer of zinc has been scratched away, so the iron continues to be protected (Figure 10.21).

▲ **Figure 10.21** The Burnley Singing Ringing Tree is a sculpture made from galvanised tubes

Sacrificial protection

Bars of zinc are attached to the hulls of ships and to oil rigs (as shown in Figure 10.22a). Zinc is above iron in the reactivity series and will react in preference to it and so is corroded. It forms positive ions more easily than the iron.

$$Zn(s) + Fe^{2+}(aq) \rightarrow Zn^{2+}(aq) + Fe(s)$$

As long as some of the zinc bars remain in contact with the iron structure, the structure will be protected from rusting. When the zinc runs out, it must be renewed. Gas and water pipes made of iron and steel are connected by a wire to blocks of magnesium to obtain the same result. In both cases, as the more reactive metal corrodes it loses electrons to the iron and so protects it (Figure 10.22b).

a Bars of zinc on the hull of a ship

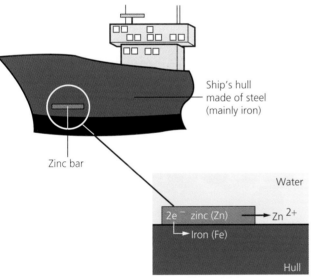

b The zinc is sacrificed to protect the steel. Electrons released from the dissolving zinc cause reduction to occur at the surface of the hull

▲ **Figure 10.22** Sacrificial protection

Corrosion

Rusting is the most common form of corrosion but this term is confined to iron and steel. **Corrosion** is the general name given to the process which takes place when metals and alloys are chemically attacked by oxygen, water or any other substances found in their immediate environment. Generally, the higher the metal is in the reactivity series, the more rapidly it will corrode. If sodium and potassium were not stored under oil they would corrode very

rapidly indeed. Magnesium, calcium and aluminium are usually covered by a thin coating of oxide after initial reaction with oxygen in the air. Freshly produced copper is pink in colour (Figure 5.15a on p. 75). However, on exposure to air, it soon turns brown due to the formation of copper(II) oxide on the surface of the metal.

In more exposed environments, copper roofs and pipes quickly become covered in verdigris. Verdigris is green in colour (Figure 10.23) and is composed of copper salts formed on copper. The composition of verdigris varies depending on the atmospheric conditions, but includes mixed copper(II) carbonate and copper(II) hydroxide ($CuCO_3.Cu(OH)_2$).

Gold and platinum are unreactive and do not corrode, even after thousands of years.

▲ **Figure 10.23** Verdigris soon forms on the surface of copper materials in exposed environments

> ### Test yourself
>
> 11 What is rust? Explain how rust forms on structures made of iron or steel.
> 12 Rusting is a redox reaction. Explain the process of rusting in terms of oxidation and reduction (Chapter 5, p. 68).
> 13 Design an experiment to help you decide whether steel rusts faster than iron.
> 14 Why do car exhausts rust faster than other structures made of steel?

10.7 Alloys

The majority of the metallic substances used today are **alloys**. Alloys are mixtures of a metal with other elements. It is generally found that alloying produces a metallic substance that has more useful properties

a   Bronze is often used in sculptures

▲ **Figure 10.24**

b   A polarised light micrograph of brass showing the distinct grain structure of this alloy

than the original pure metal it was made from. For example, the alloy brass is made from copper and zinc. The alloy is harder and more corrosion resistant than either of the metals it is made from.

Steel is a mixture of the metal iron and the non-metal carbon. Of all the alloys we use, steel is perhaps the most important. Many steel alloys have been produced; they contain iron, varying amounts of carbon, and other metals. For example, nickel and chromium are the added metals when stainless steel is produced. The chromium prevents the steel from rusting while the nickel makes it harder.

> **Key definition**
> **Alloys** can be harder or stronger than the pure metals and are more useful.

Alloys to order

Just as the properties of iron can be changed by alloying, so the same can be done with other useful metals. Metallurgists have designed alloys to suit a wide variety of different uses. Many thousands of alloys are now made, with the majority being designed to do a particular job (Figure 10.24).

Table 10.5 shows some of the more common alloys, together with some of their uses.

Many alloys are harder or stronger than the pure metals they are formed from.

In Figure 10.25 you can see that by mixing together metals which have different sized atoms to form the alloy, the metal no longer has a repeating structure. A metal is malleable and ductile because its layers are able to move over one another. In an alloy, the different sized atoms mean that the layers can no longer slide over each other, causing the alloy to have less malleability and ductility.

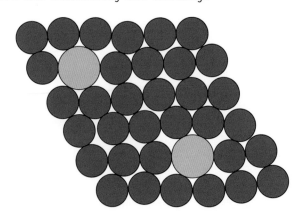

▲ **Figure 10.25** Alloy structure. The dark circles represent atoms of a metal; the pale circles are the larger atoms of a different metal added to make the alloy. The different size of these atoms gives the alloy different physical properties from those of the pure metal

▼ **Table 10.5** Uses of common alloys

Alloy	Composition	Use
Brass	65% copper, 35% zinc	Jewellery, machine bearings, electrical connections, door furniture
Bronze	90% copper, 10% tin	Castings, machine parts
Cupro-nickel	30% copper, 70% nickel	Turbine blades
Cupro-nickel	75% copper, 25% nickel	Coinage metal
Duralumin	95% aluminium, 4% copper, 1% magnesium, manganese and iron	Aircraft construction, bicycle parts
Magnalium	70% aluminium, 30% magnesium	Aircraft construction
Pewter	30% lead, 70% tin, a small amount of antimony	Plates, ornaments and drinking mugs
Solder	70% lead, 30% tin	Connecting electrical wiring

> **Test yourself**
>
> **15** 'Many metals are more useful to us when mixed with some other elements'. Discuss this statement with respect to stainless steel.

Revision checklist

After studying Chapter 10 you should be able to:
✓ State the general properties of metals.
✓ Describe the chemical reactions of metals with acids, water/steam and oxygen.
✓ Give some common uses of metals and explain why a particular metal is used in that way.
✓ Explain what an alloy is and why its properties are different to the pure metals it contains.
✓ State and explain the order of the reactivity series and use it to predict the chemical reactions of the metals.
✓ State that unreactive metals occur uncombined (not as compounds) and describe how the method used to extract reactive metals from their ores depends on their position in the reactivity series.

✓ Describe displacement experiments which could be carried out to find the order of reactivity of metals.
✓ Explain why aluminium is so unreactive.
✓ Describe how iron is extracted from its ore, hematite.
✓ State that unreactive metals occur uncombined (not as compounds) and describe how the method used to extract reactive metals from their ores depends on their position in the reactivity series.
✓ Describe methods for preventing the corrosion of metals.
✓ Explain how zinc is able to sacrificially protect iron.
✓ Explain why alloys are often harder and stronger than the metals they are composed of.

Exam-style questions

1 Use the following list of metals to answer the questions **a** to **i**: iron, calcium, potassium, gold, aluminium, magnesium, sodium, zinc, platinum.
   a Identify the metals that are found native. [1]
   b Identify which of the metals is found in nature as the ore:
      i rock salt [1]
      ii bauxite. [1]
   c Identify the metal that has a carbonate found in nature called chalk. [1]
   d Identify which of the metals will not react with oxygen to form an oxide. [1]
   e Identify which of the metals will react violently with cold water. [1]
   f Choose one of the metals in your answer to **e** and give a balanced chemical equation for the reaction which takes place. [1]
   g Identify which of the metals has a protective oxide coating on its surface. [1]
   h Identify which of the metals reacts very slowly with cold water but extremely vigorously with steam. [1]
   i Identify which of the metals is used to galvanise iron. [1]

2

Copper
↓
heat in air (oxygen)
↓
Black solid **A**
↓
heat with powdered magnesium
↓
Brown–pink solid **B** +
white powder **C**
↓
dilute hydrochloric acid
↓
Brown–pink solid **B** +
colourless solution **D** + water
↓
filter and evaporate filtrate to dryness
↓
Solid **D**
↓
electrolysis of molten **D**
↓
Silvery metal **E** + green gas **F**

a Give the name and formulae of the substances **A** to **F**. [6]
   b Give balanced chemical equations for the reactions in which:
      i black solid **A** was formed [1]
      ii white powder **C** and brown–pink solid **B** were formed [1]
      iii colourless solution **D** was formed. [1]
   c The reaction between black solid **A** and magnesium is a redox reaction. With reference to this reaction, explain what you understand by this statement. [3]
   d Write anode and cathode reactions for the processes which take place during the electrolysis of molten **D**. [2]
   e Suggest a use for:
      i brown–pink solid **B** [1]
      ii silvery metal **E** [1]
      iii green gas **F**. [1]

3 Explain the following:
   a Metals, such as gold and silver, occur native in the Earth's crust. [2]
   b The parts of shipwrecks made of iron rust more slowly in deep water. [2]
   c Zinc bars are attached to the structure of oil rigs to prevent them from rusting. [2]
   d Copper roofs quickly become covered with a green coating when exposed to the atmosphere. [2]
   e Recycling metals can save money. [2]

4 Iron is extracted from its ores hematite and magnetite. Usually it is extracted from hematite (iron(III) oxide). The ore is mixed with limestone and coke, and reduced to the metal in a blast furnace. The following is a brief outline of the reactions involved.

coke + oxygen → gas **X**

gas **X** + coke → gas **Y**

iron(III) oxide + gas **Y** → iron + gas **X**

a Name the gases **X** and **Y**. [2]
   b Give a chemical test to identify gas **X**. [2]
   c Write balanced chemical equations for the reactions shown above. [3]

d The added limestone is involved in the following reactions:

limestone → calcium oxide + gas **X**

calcium oxide + silicon(IV) oxide → slag

  i Give the chemical names for limestone and slag. [2]
  ii Write balanced chemical equations for the reactions shown above. [2]
  iii Explain why the reaction between calcium oxide and silicon(IV) oxide is called an acid–base reaction. [2]
  iv Describe what happens to the liquid iron and slag when they reach the bottom of the furnace. [2]
e Suggest why the furnace used in the extraction of iron is called a blast furnace. [1]

5 Zinc can be reacted with steam using the apparatus shown. When gas **A** is collected, mixed with air and ignited. it gives a small pop. A white solid **B** remains in the test tube when the reaction has stopped and the apparatus cooled down.

Rocksil wool
soaked in
water

B

A

Heat   Heat

Water

a Give the name and formula of gas **A**. [2]
b i Identify the product formed when gas **A** burns in air. [1]
  ii Give a balanced chemical equation for this reaction. [1]

c i Identify white solid **B**. [1]
  ii Give a balanced chemical equation to represent the reaction between zinc and steam. [1]
d Identify another metal which could be safely used to replace zinc and produce another sample of gas **A**. [1]
e When zinc reacts with dilute hydrochloric acid, gas **A** is produced again. Give a balanced chemical equation to represent this reaction and name the other product of this reaction. [2]

6 Students set up the experiment below to find out the conditions which were needed for rusting to occur.

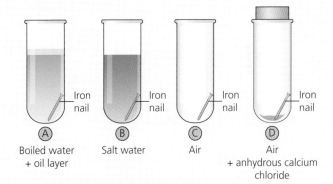

A   Boiled water + oil layer
B   Salt water
C   Air
D   Air + anhydrous calcium chloride

a Give the purpose of:
  i the anhydrous calcium chloride in tube D [1]
  ii boiling the water in tube A [1]
  iii the layer of oil in tube A. [1]
b i Identify which of the tubes contain air. [1]
  ii Identify which of the tubes contain water. [1]
  iii Identify which tube(s) contain air and water. [1]
c Identify the tube(s) in which the nails will not rust. [1]

11 Chemistry of the environment

FOCUS POINTS

★ What is the difference between pure water and water from natural sources?
★ What are the problems and advantages of the substances in water from natural sources?
★ How is water purified for domestic use?
★ Why do we use fertilisers?
★ Where do the pollutants in our air come from?
★ How can we reduce the problems of these pollutants?

In this chapter you will learn about the most plentiful liquid on this planet, **water**, and the most important gaseous mixture we know, the **air** (or atmosphere). The importance of water and air is not in doubt. Without either of these, life would not be possible on Earth! Also, we consider the problems with the pollution we have created worldwide, including excessive use of fertilisers, fossil fuels and plastics, and the greenhouse effect and the associated global warming. You will also learn about some possible solutions to the problems. By the end of this chapter you will be able to provide a possible answer to the question:

How can we guard our planet against the problems we have and those we continue to create?

11.1 Water

Water is the commonest compound on this planet. More than 70% of the Earth's surface is covered with sea, and the land masses are dotted with rivers and lakes (Figure 11.1a). It is vital to our existence and survival because it is one of the main constituents in all living organisms. For example, your bones contain 72% water, your kidneys are about 82% water and your blood is about 90% water (Figure 11.1b).

a Millions of tonnes of water pass over this waterfall every day

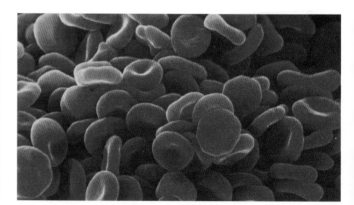

b Your blood contains a lot of water

▲ Figure 11.1

Water has many other important uses besides sustaining life. These include:

In the home:

» cooking
» cleaning
» drinking.

In industry:

» as a solvent
» as a coolant
» for cleaning
» as a chemical reactant.

Pure water is a neutral, colourless liquid which (at 1 atmosphere pressure) boils at 100°C and freezes at 0°C (Figure 11.2). If you try boiling tap water you will find that it does not boil at exactly 100°C. This is because it is a very good solvent and dissolves many substances. It is the presence of these impurities that causes a change in the boiling point compared to pure water. Distilled water is therefore used in many experiments in the laboratory because it is very pure and does not contain many of these impurities, which may affect the outcome of the experiments.

▲ **Figure 11.2** Liquid water boils at 100°C and freezes to form ice at 0°C

You can find out whether a colourless liquid contains water by adding the unknown liquid to anhydrous copper(II) sulfate. If this changes from white to blue, then the liquid contains water (Figure 11.3a).

Another test is to dip blue cobalt(II) chloride paper into the liquid. If the paper turns pink, then the liquid contains water (Figure 11.3b).

You have already seen in Chapter 5 that water may be electrolysed (when acidified with a little dilute sulfuric acid). When this is done, the ratio of the volume of the gas produced at the cathode to that produced at the anode is 2 : 1. This is what you might expect, since the formula of water is H_2O!

The unique properties of water

Water is a unique substance. Not only is it an excellent solvent for many ionic substances, such as sodium chloride, but it also has some unusual properties. For example:

» It has an unusually high boiling point for a molecule of its relatively low molecular mass.
» It has a greater specific heat capacity than almost any other liquid.
» It decreases in density when it freezes (Figure 11.4).

a Anhydrous copper(II) sulfate goes blue when water is added to it

b Cobalt(II) chloride paper turns pink when water is dropped on to it

▲ **Figure 11.3** Tests for the presence of water

▲ **Figure 11.4** When water freezes, its density falls; this is why icebergs float

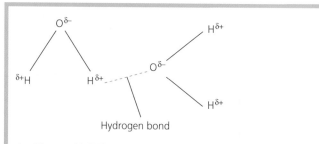

▲ **Figure 11.5** The unusual properties of water can be explained by hydrogen bonding

Water pollution and treatment

An adequate supply of fresh and safe water is essential to the health and well-being of the world's population. Across the planet, biological and chemical pollutants are affecting the quality of our water. Lack of availability of fresh water leads to waterborne diseases, such as cholera and typhoid, and to diarrhoea, which is one of the biggest killers across the world.

Agriculture needs a water supply in order to irrigate crops, especially in areas of the world with hot climates. The production of more and more crops for the ever-increasing population is essential.

> **Key definition**
>
> **Pollution** is a modification that takes place to the environment, water or air, caused by human influence such as releasing substances into our water or air.

> ➡ **Going further**
>
> These unusual properties can be explained by hydrogen bonding. This is a weak intermolecular force (bond) which occurs between water molecules because the bonds within the molecules are polar. A polar bond is one which contains, for example, oxygen attached to hydrogen by a covalent bond. The shared electrons in the bond are drawn towards the oxygen atom to create a small negative charge on the oxygen atom ($\delta-$). There is then an equally small positive charge on the hydrogen atom ($\delta+$). The water molecules then attract one another as shown in Figure 11.5. In the case of water, this attraction is called a hydrogen bond. It is a much weaker bond than a covalent bond.

Water is very good at dissolving substances and so it is very unusual to find really pure water on this planet. As water falls through the atmosphere, on to and then through the surface of the Earth, it dissolves a tremendous variety of substances including:

» a variety of gases from the air such as carbon dioxide and oxygen. However, oxygen is beneficial to aquatic life and allows it to thrive in our rivers and oceans. Carbon dioxide creates a low-level natural acidity in the water, but this acidity is increasing, due to the increase in the amount of carbon dioxide released into the atmosphere by vehicles and industry.

» nitrates and phosphates from agricultural waste and detergents. Chemical fertilisers, washed off surrounding land, add nitrate ions (NO_3^-) and phosphate ions (PO_4^{3-}) to the water. This is due to the use of artificial fertilisers, such as ammonium nitrate and ammonium phosphate (see below), as well as some pesticides. Detergents used in the home and in industry contain phosphates. The nitrates and phosphates encourage the growth of algae which eventually die and decay, removing oxygen from the water (deoxygenation). They can also disrupt sensitive ecosystems.

» metal compounds from industrial waste water may contain harmful chemicals, such as cadmium and mercury, which are **toxic**. It should be noted however, that not all metal compounds are toxic. Some metal compounds that are present are also beneficial as they provide essential minerals. This is especially true of calcium-containing compounds. The calcium present in water is necessary for healthy growth of bones and teeth.

» human waste from sewage. Sewage contains harmful microbes which cause disease.

» insoluble impurities such as oil and plastic waste (Figure 11.6). Plastic waste is polluting our streams and rivers as well as oceans. It is detrimental to all aquatic life as well as human life on this planet. This issue will be discussed in more detail in Chapter 12.

▲ **Figure 11.6** A badly polluted river

All these artificial and natural impurities must be removed from the water before it can be used. Recent regulations in many countries have imposed strict guidelines on the amounts of various substances allowed in drinking water.

▲ **Figure 11.7** This lake is used as a source of drinking water

A lot of drinking water is obtained from lakes and rivers where the **pollution** levels are low (Figure 11.7). Undesirable materials are removed from water by the process of water treatment, which involves both filtration and chlorination and is summarised in Figure 11.8.

1 Impure water is passed through screens to filter out floating debris.

2 Aluminium sulfate is added to coagulate small particles of clay so that they form larger clumps, which settle more rapidly.

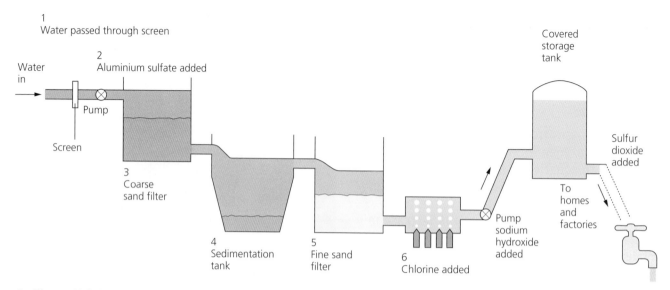

▲ **Figure 11.8** The processes involved in water treatment

3 Filtration through coarse sand traps larger, insoluble particles. The sand also contains specially grown microbes which remove some of the bacteria.

4 A sedimentation tank has chemicals known as flocculants, for example aluminium sulfate, added to it to make the smaller particles (which remain in the water as colloidal clay) stick together and sink to the bottom of the tank.

5 These particles are removed by further filtration through fine sand. This is followed by carbon slurry filters which are there to remove unwanted tastes and odours, and a lime slurry is used to adjust the acidity.

6 Finally, a little chlorine gas is added, which sterilises the water and kills any remaining bacteria. Excess chlorine can be removed by the addition of sulfur dioxide gas. The addition of chlorine gas makes the water more acidic and so appropriate amounts of sodium hydroxide solution are added. Fluoride is sometimes added to water if there is insufficient occurring naturally, as it helps to prevent tooth decay.

11.2 Artificial fertilisers

Some of the ammonia produced by the Haber process (see Chapter 7, p. 105) is used to produce nitric acid. If ammonia is then reacted with the nitric acid, ammonium nitrate is produced. This gives us the basic reaction for the production of many artificial **fertilisers**.

> **Key definition**
>
> **Fertiliser** is a chemical put onto soil to replace lost mineral salts and so make plants grow more healthily. These include ammonium salts, such as ammonium nitrate, which is one of the most commonly used fertilisers.

ammonia + nitric acid → ammonium nitrate

$$NH_3(g) + HNO_3(aq) \rightarrow NH_4NO_3(aq)$$

Ammonium nitrate (Nitram®) is probably the most widely used nitrogenous fertiliser. The use of artificial fertilisers is essential if farmers are to produce sufficient crops to feed the ever-increasing world population. Crops remove nutrients from the soil as they grow; these include nitrogen, phosphorus and potassium. Artificial fertilisers are added to the soil to replace these nutrients and others, such as calcium, magnesium, sodium, sulfur, copper and iron. Examples of nitrogenous fertilisers (those which contain nitrogen) are shown in Table 11.1.

▼ **Table 11.1** Some nitrogenous fertilisers

Fertiliser	Formula
Ammonium nitrate	NH_4NO_3
Ammonium phosphate	$(NH_4)_3PO_4$
Ammonium sulfate	$(NH_4)_2SO_4$
Urea	$CO(NH_2)_2$

Artificial fertilisers can also create fertile land from areas unable to support crop growth. The fertilisers which add the three main nutrients (N, P and K) are called NPK fertilisers. They contain ammonium nitrate (NH_4NO_3), ammonium phosphate (($NH_4)_3PO_4$) and potassium chloride (KCl) in varying proportions (Figure 11.9). Fertilisers have an important role in the nitrogen cycle.

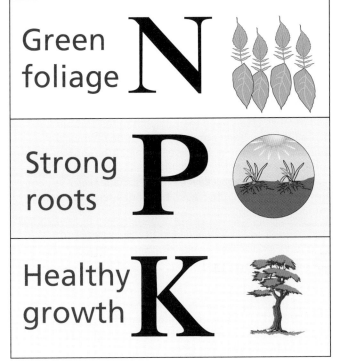

a Different fertilisers contain differing amounts of the elements nitrogen, phosphorus and potassium.

b The different NPK elements are responsible for the healthy growth of plants in different ways.

▲ **Figure 11.9**

▲ **Figure 11.10** Fertilisers have been used to help create some of the best fruit and vegetables on sale

❓ Worked example

Using relative atomic masses and the formula of a fertiliser, it is possible to calculate the percentage of each of the essential elements in a fertiliser.

What is the percentage of nitrogen in the fertiliser ammonium sulfate ($(NH_4)_2SO_4$)?

(A_r: H = 1; N = 14; O = 16; S = 32)

The relative formula mass of ammonium sulfate ($(NH_4)_2SO_4$) is:

$(2 \times 14) + 2(4 \times 1) + 32 + (4 \times 16) = 132$

The % nitrogen in ammonium sulfate is:

$((2 \times 14)/132) \times 100 = 21.2\%$

Problems with fertilisers

If artificial fertilisers of all kinds are not used correctly, problems can arise. If too much fertiliser is applied to the land, rain washes the fertiliser off the land and into rivers and streams. This is known as leaching. This leaching leads to **eutrophication**: the process that occurs when fertiliser is leached, causing algae to multiply rapidly and causing the water to turn green. As the algae die and decay, oxygen is removed from the water, leaving insufficient amounts for fish and other organisms to survive (Figure 11.12). In extreme cases, no normal aquatic life can survive. There are also worries about the effect of agricultural fertilisers, especially nitrates such as ammonium nitrate, on the public water supply.

Going further

The vital importance of nitrogen to both plants and animals can be summarised by the nitrogen cycle (Figure 11.11).

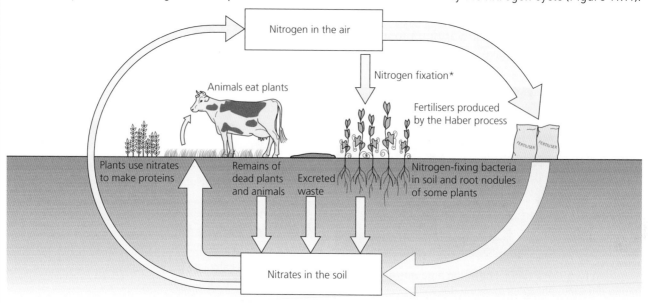

▲ **Figure 11.11** The nitrogen cycle
*Nitrogen fixation is the direct use of atmospheric nitrogen in the formation of important compounds of nitrogen. Bacteria present in the root nodules of certain plants are able to take nitrogen directly from the atmosphere to form essential protein molecules.

If farm crops are harvested from the land rather than left to decay, the soil becomes deficient in this important element. The nitrogen is removed in the harvested crops rather than remaining as the plants decay. In addition, nitrates can be washed from the soil by the action of rain (leaching). For the soil to remain fertile for the next crop, the nitrates need to be replaced. The natural process is by decay or by the action of lightning on atmospheric nitrogen. Without the decay, however, the latter process is not efficient enough to produce nitrates on the scale required.

Farmers often need to add substances containing these nitrates. Such substances include farmyard manure and artificial fertilisers. One of the most commonly used artificial fertilisers is ammonium nitrate, which, as you saw earlier, is made from ammonia gas and nitric acid, both nitrogen-containing compounds.

▲ **Figure 11.12** Over-use of fertilisers has led to eutrophication in this river

Test yourself

1 Calculate the percentage of nitrogen in each of the three fertilisers ammonium nitrate (NH_4NO_3), ammonium phosphate ((NH_4)$_3PO_4$) and urea $CO(NH_2)_2$ (A_r: H = 1; C = 12; N = 14; O = 16; P = 31)
2 Write down a method that you could carry out in a school laboratory to prepare a sample of ammonium sulfate fertiliser.
3 Why is it important to have nitrogen in fertilisers?

11.3 The air

The gases in the air are held around the Earth by its gravity. The atmosphere is approximately 100 km thick (Figure 11.13), and about 75% of the mass of the atmosphere is found in the layer nearest the Earth called the troposphere (Figure 11.14). Beyond this layer, the atmosphere reaches into space but becomes extremely thin. Nearly all atmospheric water vapour (or moisture) is found in the troposphere, which also contains the liquid water in the oceans, rivers and lakes.

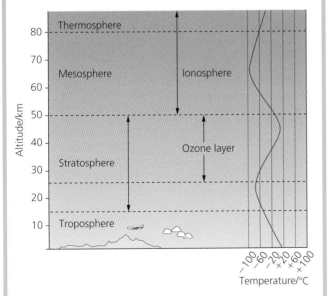

▲ **Figure 11.13** The lighter blue shows the extent of the atmosphere around the Earth – approximately 100 km

▲ **Figure 11.14** The Earth's atmosphere

The composition of the atmosphere

If a sample of dry, unpolluted air was taken from any location in the atmosphere close to the Earth and analysed, the composition by volume of the sample would be similar to that shown in Table 11.2.

▼ **Table 11.2** Composition of the atmosphere

Component	%
Nitrogen	78.08
Oxygen	20.95
Argon	0.93
Carbon dioxide	0.04
Neon	0.002
Helium	0.000 5
Krypton	0.000 1
Xenon plus tiny amounts of other gases	0.000 01

> **Key definition**
>
> Clean, dry air is approximately 78% nitrogen, 21% oxygen and the remainder is a mixture of noble gases and carbon dioxide.

The oxygen in the atmosphere is produced by plants. Carbon dioxide is taken in by plants through their leaves and used together with water, taken in through their roots, to synthesise glucose (a sugar). Oxygen is also produced. This is the process of **photosynthesis**.

It takes place only in sunlight and only in green leaves, as they contain chlorophyll (the green pigment) which catalyses the process.

$$\text{carbon dioxide} + \text{water} \xrightarrow[\text{chlorophyll}]{\text{sunlight}} \text{glucose} + \text{oxygen}$$

$$6CO_2 + 6H_2O \rightarrow C_6H_{12}O_6 + 6O_2$$

Note that air contained approximately 0.03% by volume of carbon dioxide for many years. This value has remained almost constant for a long period of time. However, scientists have recently detected an increase in the amount of carbon dioxide in the atmosphere to approximately 0.04%.

The Earth's climate is affected by the levels of carbon dioxide (and water vapour) in the atmosphere. If the amount of carbon dioxide, in particular, builds up in the air, it is thought that the average temperature of the Earth will rise causing **global warming**. This effect is thought by scientists to be caused by the **greenhouse effect**. This important topic will be discussed later in this chapter, on p. 176.

> ### Test yourself
>
> 4 Draw a pie chart to show the data given in Table 11.2.
> 5 Is air a compound or a mixture? Explain your answer.
> 6 Which of the following statements about the air are true and which are false?
>   a Nitrogen constitutes nearly 79% of the atmosphere.
>   b Air is a mixture of elements.
>   c Carbon dioxide is not a constituent of the air.
>   d Neon is present in the air.

11.4 Atmospheric pollution

The two major resources considered in this chapter, water and air, are essential to our way of life and our very existence. Water and air make up the environment of a living organism. The environment is everything in the surroundings of an organism that could possibly influence it. Humans continually pollute these resources. We now look at the effects of the various sources of pollution of the air and at the methods used to control or eliminate them. For a discussion of water pollution, see p. 166.

Air pollution is all around us. Concentrations of gases in the atmosphere, such as carbon monoxide, sulfur dioxide and nitrogen oxides, are increasing with the increasing population. As the population increases, there is an increase in the need for energy, industries and motor vehicles. These gases are produced mostly from the combustion of the fossil fuels, coal, oil and gas, but they are also produced by the smoking of cigarettes.

Motor vehicles are responsible for much of the air pollution in large towns and cities. They produce five particularly harmful pollutants, as shown in Table 11.3.

▼ **Table 11.3** Pollution caused by motor vehicles

Pollutant	Caused by	Problems caused	Can be reduced by
Carbon monoxide, CO	The incomplete combustion of carbon-containing fuels	Toxic gas	Fitting catalytic converters to remove CO, but more CO_2 is produced. It should be noted that lean-burn engines also produce less CO
Particulates, C	The incomplete combustion of carbon-containing fuels	Increased risk of respiratory problems and cancer	Adding oxygenates such as ethanol to petrol to reduce the particulates produced
Methane, CH_4	The decomposition of vegetation and waste gases from digestion in animals	Higher levels of methane lead to increased global warming, which leads to climate change	Encouraging composting or incineration. Capturing the methane produced and using it as a fuel or to generate electricity when burned. Reducing the number of farm animals by encouraging people to eat less meat
Nitrogen oxides, NO_x	Car engines	Acid rain, photochemical smog and respiratory problems	Fitting catalytic converters to remove nitrogen oxides. However, in doing so more carbon dioxide is produced
Sulfur dioxide, SO_2	The combustion of fossil fuels which contain sulfur compounds	Acid rain	Removing sulfur from petrol to produce low-sulfur petrol. Using flue gas desulfurisation units (see p. 175) at coal-burning power stations

Catalytic converters

Many countries have 'clean air' laws that all new petrol (gasoline) cars have to be fitted with **catalytic converters** as part of their exhaust system (Figure 11.15). Car exhaust fumes contain pollutant gases, such as carbon monoxide (CO), formed from the incomplete combustion of hydrocarbons (C_xH_y) in the fuel, and oxides of nitrogen (NO_x) formed by the reaction at very high temperature of nitrogen gas and oxygen gas from the air. Nitrogen oxides in vehicle exhausts include nitrogen(II) oxide (NO) and nitrogen(IV) oxide (NO_2).

The following reaction happens naturally inside a car exhaust but is very slow under the conditions inside an exhaust. The catalyst in the catalytic converter speeds up the reaction by lowering the activation energy.

nitrogen(II) + carbon → nitrogen + carbon
oxide   monoxide   dioxide

$$2NO(g) + 2CO(g) \rightarrow N_2(g) + 2CO_2(g)$$

Carbon monoxide is toxic, so the reaction shown here removes two pollutant gases. The pollutants are converted to carbon dioxide and nitrogen, which are naturally present in the air. Other reactions catalysed in the converter remove NO_2. The catalyst is made from platinum or other transition metals, such as palladium or rhodium. The removal of oxides of nitrogen is important because they cause respiratory disease. They are also involved in the production of photochemical **smogs** (Figure 11.16) which occur worldwide in major cities, especially in the summer. Photochemical smog is the most widely known and perhaps most serious air pollutant. It is formed in the atmosphere by the reaction between gaseous pollutants, nitrogen oxides and hydrocarbons.

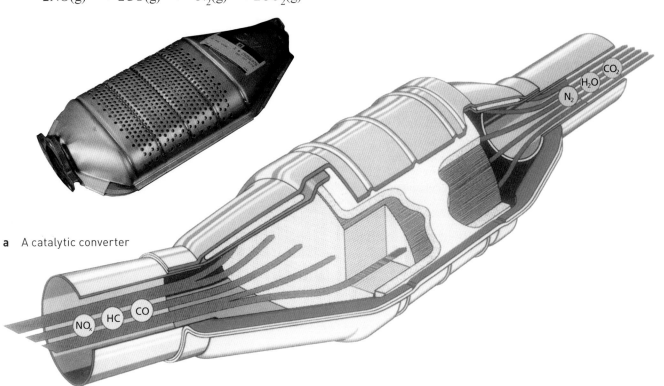

a   A catalytic converter

b   A section through a catalytic converter

▲ **Figure 11.15** A catalytic converter removes hydrocarbons, carbon monoxide and oxides of nitrogen from vehicle exhausts

▲ **Figure 11.16** The haze is caused by a photochemical smog, which is due to pollution caused mainly by cars without catalytic exhaust systems

> ### Test yourself
>
> 7 Catalytic converters remove harmful gases CO and NO. They do this by converting them into less harmful chemicals. Give the names and formulas of the chemicals they are converted to.

 Going further

The incomplete combustion of hydrocarbons in the petrol and the evaporation of hydrocarbons from petrol tanks both contribute to the formation of photochemical smog. This can cause severe respiratory problems. However adding oxygenates, such as ethanol, to petrol reduces its production. Catalytic converters also help remove unburned hydrocarbons but in doing so more carbon dioxide is produced!

A further method of regulating pollutant gases from vehicles is to remove petrol-burning engines and replace them with an efficient electric motor. The research and development of these motors is moving forward at a tremendous pace.

Particulates

Internationally there has been great concern globally with the increase of **particulates** in the air (Figure 11.17). 'Particulates' is a general term used to describe very small particles in the atmosphere, such as certain types of smoke emitted from diesel engines, as well as dust. Particulates in the smoke emitted from vehicle engines and also from burning fuels such as wood or coal are produced by the incomplete combustion of the fuel.

 Going further

These particulates have been associated with a variety of breathing problems, including asthma, in both adults and children. It has been found, however, that adding oxygenates, such as ethanol, to fuels reduces the amount of particulates produced by vehicles. The oxygenates provide extra oxygen to the burning process in the engine and so reduce the amount of incomplete combustion taking place.

▲ **Figure 11.17** Particulates produced by diesel engines are a real health problem worldwide

Acid rain

In addition to the fuels used in motor vehicles producing quantities of sulfur dioxide when they combust in engines, heavy industry (Figure 11.18) and power stations are also major sources of sulfur dioxide, formed by the combustion of coal, oil and gas, which also contain small amounts of sulfur.

▲ **Figure 11.19** This forest has been devastated by acid rain

▲ **Figure 11.18** Sulfur dioxide is a major pollutant produced by industry

Rainwater is naturally acidic since it dissolves carbon dioxide gas from the atmosphere as it falls. Natural rainwater has a pH of about 5.7. However, the acidity is enhanced by sulfur dioxide gas dissolving in rainwater to form sulfurous acid (H_2SO_3).

A further reaction occurs in which the sulfurous acid is oxidised to sulfuric acid (H_2SO_4). Solutions of these acids are the principal contributors to acid rain.

Vehicle exhaust fumes contain another pollutant gas that causes **acid rain**. Nitrogen(IV) dioxide (NO_2) is another oxide of nitrogen produced inside vehicle engines. It dissolves in rainwater and produces, eventually, nitric acid.

First noticed as a problem in Europe and North America, acid rain is a growing concern in parts of Asia. China and India's sulfur dioxide emissions are the highest in the world, and this affects the pH of rainwater in many parts of the region. Acid rain can damage buildings by the reaction with metals, concrete and stone. A decrease in the pH of rainwater also makes soils and lakes become acidic, which damages trees and ecosystems (Figure 11.19).

A catalytic converter removes nitrogen oxides from the hot gases coming from the engine and so helps reduce acid rain. Recently, units called flue gas desulfurisation (FGD) units have been fitted to some fossil fuel power stations throughout the world to prevent the emission of sulfur dioxide gas. Here, the sulfur dioxide gas is removed from the waste gases by passing them through a mixture of calcium oxide suspended in water. This not only removes the sulfur dioxide but also creates calcium sulfite, which then oxidises to sulfate, which can be sold to produce plasterboard (Figure 11.20). The FGD units are very expensive and therefore the sale of the calcium sulfate is an important economic part of the process.

▲ **Figure 11.20** This plasterboard is made using calcium sulfate from an FGD plant

Practical skills

Acid rain

Safety

● Eye protection must be worn.

Experiment A

A student tests the rainwater collected from the school playing field using universal indicator solution. They then tested rainwater from three other places, A, B and C.

1 What apparatus would be required for this experiment?
The results of the tests are shown in the table below.

Sample	Colour of universal indicator solution	pH of rain water
Playing fields	Lime green	6
A	Orange	3
B	Yellow	5
C	Lime green	6

2 Which was the most acidic water?
3 Which was the least acidic water?

Experiment B

The student was then given some powdered calcium carbonate to add to each sample. The results are shown in the table below. The student was asked to collect samples of any gas produced and test any gas produced using limewater.

Sample	Observation upon adding calcium carbonate powder	Observation after adding limewater
Playing Fields	Small amount of bubbles but not sufficient to test	No result
A	Lots of bubbles of a gas produced	Limewater turns milky white
B	Small amount of bubbles but not sufficient to test	No result
C	Small amount of bubbles but not sufficient to test	No result

4 a Which gas was being tested for?
  b How could a sample of the white solid be obtained after the test was carried out on A?
5 Describe, with the aid of a simple diagram, how the student could collect samples of any gas produced.
6 Did these results confirm the results obtained in experiment A? Explain your answer.
7 Write a conclusion to experiments A and B.

Global warming

Scientists know that certain gases in the atmosphere trap the Sun's heat in our atmosphere and this makes the Earth warmer. These gases are known as greenhouse gases and cause the greenhouse effect. This increase in the Earth's temperature is known as global warming.

It is well known by scientists that, for example, higher levels of carbon dioxide in the atmosphere lead to increased global warming, which leads to climate change.

How does the greenhouse effect affect the average temperature of the Earth?

Some energy from the Sun is absorbed by the Earth and its atmosphere. The remainder is reflected back into space. The energy that is absorbed helps to heat up the Earth. The Earth radiates some thermal energy back into space but the 'greenhouse gases', including carbon dioxide, prevent it from escaping. This effect is similar to that observed in a greenhouse (glasshouse) where sunlight (visible/ultraviolet radiation) enters through the glass panes but heat (infrared radiation) has difficulty escaping through the glass (Figure 11.21).

Carbon dioxide is not the only greenhouse gas. Other gases also contribute to the greenhouse effect. One of these is methane (see Chapter 12, p. 187), which is produced from agriculture and released from landfill sites as well as rice fields. The amount of carbon dioxide in the atmosphere is increasing for three reasons:

» Major deforestation taking place in several countries of the world. Trees act as the lungs of the planet. During photosynthesis they absorb carbon dioxide and release oxygen.

Pollution of our rivers and oceans. The more pollutants that are dissolving in these systems, the less carbon dioxide that can dissolve in the water.

Large increase in the amounts of carbon dioxide released by the continued burning of large quantities of fossil fuels by industry and transport.

The long-term effect of the higher temperatures of the greenhouse effect and the subsequent global warming will be the continued gradual melting of ice caps and consequent flooding in low-lying areas of the Earth. There will also be further changes in the weather patterns which will further affect agriculture worldwide.

There is an additional effect when the incomplete combustion of diesel takes place. Carbon particles or particulates are produced. These black particles are blown by the prevailing winds and some end up falling on snow and ice fields. This reduces the reflectivity of the snow, which causes a further increase in the melting of the snow and ice in mountainous areas as well as the northern and southern pole areas of the world.

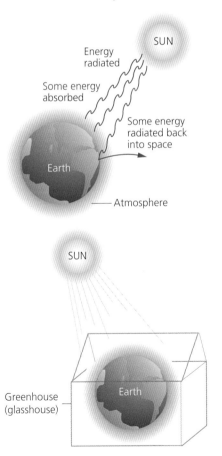

▲ **Figure 11.21** The greenhouse effect

These problems have been recognised by nations worldwide. Recent agreements under the Kyoto Protocol, as well as the Paris Accord, between nations mean that there will be some reduction in the amount of carbon dioxide (and other greenhouse gases) produced over the next few years. However, there is still a long way to go before we no longer increase greenhouse gases in the atmosphere.

How can we reduce the effects of global warming?

There are several actions that can be taken to reduce the effects of global warming:

Plant more trees and reduce the amount of deforestation taking place worldwide. This will cause more carbon dioxide to be absorbed from the atmosphere through photosynthesis.

Burn fewer fossil fuels by using alternative fuels in order to reduce the amount of carbon dioxide entering the atmosphere. You will see in Table 11.4 that there are alternatives that might improve matters. However, we must take into account whether or not the alternative fuel is going to use up other resources in a way that enables them to continue in the future rather than destroying them. In other words, is the alternative sustainable or not? Sustainability is very important. In the case of fossil fuels, they are not renewable or sustainable. When they are gone they are gone!
Also we must consider the advantages and disadvantages of the alternative fuel. Table 11.4 shows not only whether those alternatives are sustainable but also their perceived advantages and disadvantages.

Be more energy efficient. This will reduce the amount of carbon dioxide entering the atmosphere. In homes, for example, this means making them well insulated and turning off lights when not needed. We also need to consider whether the journey we are about to take by car or plane is essential. Only by being conscientious will we make inroads into the global warming situation.

Increase the use of renewable energy sources such as solar, wind and tidal.

Significantly reduce livestock farming to cut down the amount of methane released into the atmosphere.

▼ **Table 11.4** Alternative fuels to fossil fuels

Alternative fuel	Is the alternative sustainable?	Advantages	Disadvantages
Biodiesel	Yes it is! It is made from waste plant material and animal oils and fats.	Produces less CO, C_xH_y, SO_2 and particulates than diesel fuel.	NO emissions are higher than from standard diesel fuel.
Ethanol	This is open to debate since large amounts of energy and land are needed to cultivate sugar cane for fermentation.	Less CO, SO_2 and NO_x are produced than from petrol. Replanting sugar cane creates a cycle as it absorbs CO_2 from the atmosphere.	Very flammable.
Hydrogen	Hydrogen is sustainable only if the electricity needed to produce it, from the electrolysis of acidified water, is from a renewable resource such as solar power or wind. The car shown in Figure 11.22 is powered by hydrogen.	Water is the only product of production. Hence there is no pollution.	It is very, very flammable. Also a high-pressure fuel tank is needed to store it as a liquid.

▲ **Figure 11.22** A hydrogen-powered vehicle

Test yourself

8 In 1960 the percentage of carbon dioxide in the atmosphere was close to 0.03% and in 2010 it was close to 0.04%. Suggest an explanation for this increase.

9 Write a balanced chemical equation to represent the reaction which takes place between sulfur dioxide and calcium hydroxide slurry in the FGD unit of a power station.

10 Write down one problem that can be caused by each of these air pollutants:
   a nitrogen dioxide
   b particulates.

Revision checklist

After studying Chapter 11 you should be able to:
- ✔ Name the gases in the atmosphere and state the approximate percentage proportions of each gas.
- ✔ Name the gases in the atmosphere and state the approximate percentage proportions of each gas.
- ✔ Describe chemical tests for the presence of water.
- ✔ Describe how to test for the purity of water.
- ✔ Explain why distilled water is used in practical chemistry rather than tap water.
- ✔ State that ammonium salts and nitrates can be used as fertilisers, including ammonium phosphate and potassium nitrate.
- ✔ Describe the use of NPK fertilisers to provide improved plant growth and provide larger yields of crops to feed the world's growing population.
- ✔ State the composition of clean, dry air.

- ✔ State the source of these air pollutants: carbon dioxide, carbon monoxide, methane, nitrogen oxides, sulfur dioxide and particulates.
- ✔ State the harm these air pollutants cause to health or to the environment.
- ✔ Describe how the greenhouse gases, such as carbon dioxide and methane, cause global warming, which leads to climate change.
- ✔ Explain how oxides of nitrogen form in car engines and describe their removal by catalytic converters.
- ✔ State and explain strategies to reduce the effects of these various environmental issues.
- ✔ Describe photosynthesis as the reaction between carbon dioxide and water to produce glucose in the presence of chlorophyll and using energy from light.

Exam-style questions

1 The apparatus shown below was used to estimate the proportion of oxygen in the atmosphere.

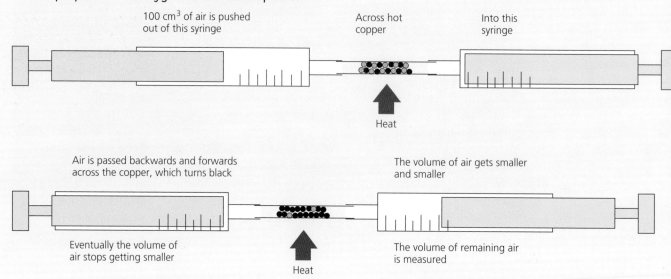

100 cm³ of air is pushed out of this syringe

Across hot copper

Into this syringe

Heat

Air is passed backwards and forwards across the copper, which turns black

The volume of air gets smaller and smaller

Eventually the volume of air stops getting smaller

Heat

The volume of remaining air is measured

A volume of dry air (200 cm³) was passed backwards and forwards over heated copper until no further change in volume took place. The apparatus was then allowed to cool down to room temperature and the final volume reading was then taken. Some typical results are shown below.

Volume of gas before = 200 cm³

Volume of gas after = 157 cm³

During the experiment the copper slowly turned black.

a Explain why the apparatus was allowed to cool back to room temperature before the final volume reading was taken. [2]

b Using the information given above, calculate the percentage volume reduction which has taken place. [3]

c Explain briefly why there is a change in volume. [2]

d Identify which observation given above supports your explanation in **c**.

e Give a balanced chemical equation for any reaction which has occurred. [4]

f Give the name of the main residual gas at the end of the experiment. [1]

g Would you expect the copper to have increased or decreased in mass during the experiment? Explain your answer. [2]

2 a Oxygen has an atomic number of 8 and mass number 16. Give the electronic configuration of the oxygen atom. [1]

b How many electrons, neutrons and protons are there in the oxygen atom? [3]

c Oxygen molecules are diatomic. Explain the meaning of this term. [3]

d Sketch a diagram of the oxygen molecule showing the outer shell of electrons only. What type of bonding does the molecule contain? [5]

3 Explain the following:

a Air is a mixture of elements and compounds. [2]

b The percentage of carbon dioxide in the atmosphere does not significantly vary from 0.04%. [2]

c Power stations are thought to be a major cause of acid rain. [3]

4 Use the words below to complete the following passage about water and its uses.

abundant chlorine monitored coolant reactions pure clay solvent essential

Water is _____ for all life on Earth. It is the most _____ substance on Earth. Water is not only used for drinking and washing but also:
● in chemical _____, for example in the production of ethanol from ethene
● as a _____ to dissolve things
● as a _____ in some chemical processes.

For us to drink it, water must be relatively _____. To make the water drinkable, any undissolved solids are removed through filtration beds. Aluminium sulfate is added to remove small particles of _____ and _____ is added to kill bacteria.

Water is continuously _____ because there are certain substances that eventually find their way into the water supply. [9]

5 a Explain what is meant by the term 'pollution' with reference to air and water. [4]
  b i Give an air pollutant produced by the burning of coal. [1]
    ii Give a different air pollutant produced by the combustion of petrol in a car engine. [1]
  c Some of our drinking water is obtained by purifying river water.

i Would distillation or filtration produce the purest water from river water? Give a reason for your answer. [3]
ii Which process, distillation or filtration, is actually used to produce drinking water from river water? Compare your answer with your answer in **c i**. [2]

6 a Besides carbon dioxide, name one other greenhouse gas. [1]
  b Carbon dioxide is a greenhouse gas which adds to the greenhouse effect. Describe what you understand by the term 'greenhouse effect'. [3]
  c The graph below shows the change in mean global air temperature from 1860 to 2000.

i Describe the general trend shown by the graph from 1860. [1]
ii Explain what you think has caused the trend you gave as an answer in part **a**. [2]

12 Organic chemistry 1

You will have seen earlier in Chapter 6 that important substances, called hydrocarbons, are obtained from petroleum. These hydrocarbons are mainly used as fuels of different types. In this chapter you will study two families, or homologous series, of these hydrocarbons, the alkanes and alkenes. You will examine their different structures and different physical and chemical properties. You will see that it is possible to produce alkenes from alkanes in a process called cracking. Finally, you will learn about a very important set of materials that are made from alkenes called polymers, better known as plastics, and the problems that are created by overuse of these polymers.

A lot of the compounds that are present in living things have been found to be compounds containing carbon (Figure 12.1). These are known as **organic compounds**. All living things are made from organic compounds based on chains of carbon atoms which are not only covalently bonded to each other but also covalently bonded to hydrogen, oxygen and/or other elements. The organic compounds are many and varied. Some scientists suggest that there are more than ten million known organic compounds.

▲ **Figure 12.1** Living things contain organic compounds

12.1 Alkanes

Most of the hydrocarbons in petroleum belong to the family of compounds called **alkanes**. The molecules within the alkane family contain carbon atoms covalently bonded to four other atoms by single bonds as shown in Figure 12.2, which shows the fully **displayed formulae** of some alkanes. A **displayed formula** shows how the various atoms are bonded and shows all the bonds in the molecule as individual lines. Because these molecules possess only carbon–carbon single covalent bonds, they are said to be **saturated**, as no further atoms can be added. This can be seen in the bonding scheme for methane (Figure 12.3). The physical properties of the first six members of the alkane family are shown in Table 12.1, p. 184.

> **Key definition**
>
> The **displayed formula** of a molecule is a diagram that shows how the various atoms are bonded and shows all the bonds in the molecule as individual lines.
>
> A **saturated** compound has molecules in which all carbon–carbon bonds are single bonds.

You will notice from Figure 12.2 and Table 12.1 that the compounds have a similar structure and similar name endings. They also behave chemically in a similar way. A family with these factors in common is called a **homologous series**.

Methane

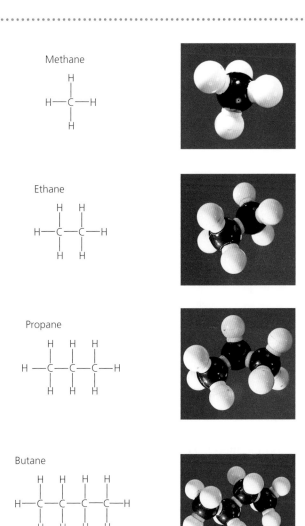

Ethane

Propane

Butane

Pentane

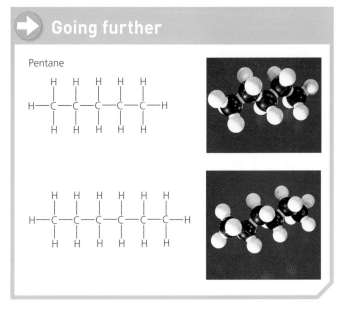

▲ **Figure 12.2** The fully displayed formulae and molecular models of the first six alkanes

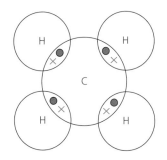

Methane molecule (CH$_4$)

▲ **Figure 12.3** The covalent bonding scheme for methane

All the members of a homologous series can also be represented by a general formula. In the case of the alkanes, the general formula is:

$$C_nH_{2n+2}$$

where *n* is the number of carbon atoms present.

As you go up a homologous series, in order of increasing number of carbon atoms, the physical properties of the compounds gradually change. For example, the melting and boiling points of the alkanes shown in Table 12.1 gradually increase. This is due to an increase in the intermolecular forces as the size and mass of the molecule increases (Chapter 3, p. 42). As you can see from Figure 12.2 and Table 12.1, the increase in size and mass of the molecule is due to the addition of a CH$_2$ group as you descend the homologous series.

The compounds within a homologous series possess similar chemical properties through their **functional group**. A functional group is usually an atom or group of atoms that are present within the molecules of the homologous series. In the case of the alkanes, they are unusual in that they do not have a functional group. For a further discussion of functional groups see Chapter 13, p. 199.

Under normal conditions, molecules with up to four carbon atoms are gases, those with between five and 16 carbon atoms are liquids, while those with more than 16 carbon atoms are solids.

Key definitions

A **homologous series** is a family of similar compounds with similar chemical properties due to the presence of the same functional group.

A **functional group** is an atom or group of atoms that determine the chemical properties of a homologous series.

▼ **Table 12.1** Some alkanes and their physical properties

Alkane	Molecular formula	Melting point/°C	Boiling point/°C	Physical state at room temperature
Methane	CH_4	–182	–162	Gas
Ethane	C_2H_6	–183	–89	Gas
Propane	C_3H_8	–188	–42	Gas
Butane	C_4H_{10}	–138	0	Gas

Going further

Alkane	Molecular formula	Melting point/ °C	Boiling point/ °C	Physical state at room temperature
Pentane	C_5H_{12}	–130	36	Liquid
Hexane	C_6H_{14}	–95	69	Liquid

Test yourself

1 Estimate the boiling points for the alkanes with formulae:
   a C_7H_{16}
   b C_8H_{18}
2 Name the alkane that has the formula C_7H_{16}.

Naming the alkanes

All the alkanes have names ending in *-ane*. The rest of the name tells you the number of carbon atoms present in the molecule. For example, the compound whose name begins with:

» *meth-* has one carbon atom
» *eth-* has two carbon atoms
» *prop-* has three carbon atoms
» *but-* has four carbon atoms

   and so on.

Structural isomerism

Sometimes it is possible to write more than one displayed formula to represent a molecular formula. The displayed formula of a compound shows how the atoms are joined together by the covalent bonds. For example, there are two different compounds

with the molecular formula C_4H_{10}. The displayed and structural formulae of these two substances, along with their names and physical properties, are shown in Figure 12.4.

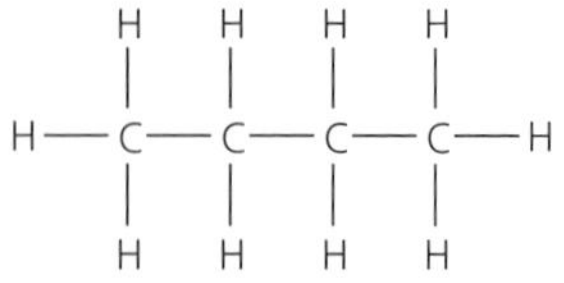

Melting point –138°C

Boiling point 0°C

$CH_3CH_2CH_2CH_3$

a   Butane

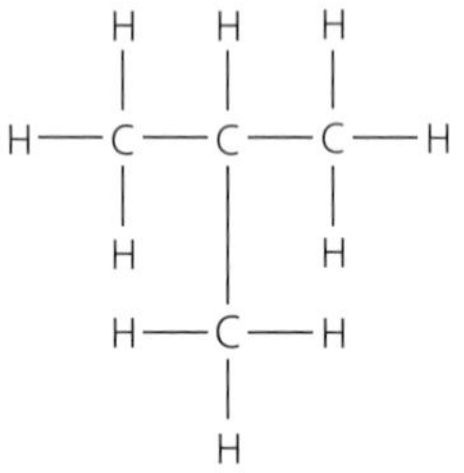

Melting point –159°C

Boiling point –12°C

$CH_3CH(CH_3)CH_3$

b   2-Methylpropane

▲ **Figure 12.4** Displayed and structural formulae for the two isomers of C_4H_{10}

Compounds such as those in Figure 12.4 are known as **structural isomers**. **Isomers** are substances which have the same molecular formula but different **structural formulae** and displayed formulae. The isomers in Figure 12.4 have the same molecular formula, C_4H_{10}, but different structural and displayed formulae. Butane (Figure 12.4a) has the structural formula $CH_3CH_2CH_2CH_3$ and 2-methylpropane (Figure 12.4b) has the structural formula $CH_3CH(CH_3)CH_3$. The brackets show the formula of the branch. The different structures of the compounds shown in Figure 12.4 have different melting and boiling points. Molecule **b** contains a branched chain and has a lower melting point than molecule **a**, which has no branched chain. All the alkane molecules with four or more carbon atoms possess isomers. Perhaps now you can see why there are so many different organic compounds!

Key definitions

Structural isomers are compounds with the same molecular formula, but different structural formulae.

The **structural formula** of an organic compound is an unambiguous description of the way the atoms are arranged, including the functional group.

> **Test yourself**
>
> 3 Draw the displayed formulae for the isomers of C_5H_{12}.

For further insight into isomerism see p. 188 and Chapter 13, p. 200.

12.2 The chemical behaviour of alkanes

Alkanes are rather unreactive compounds. For example, they are generally not affected by alkalis, acids or many other substances. Their most important property is that they burn or combust easily.

Gaseous alkanes, such as methane, will burn in a good supply of air, forming carbon dioxide and water as well as plenty of thermal energy.

methane + oxygen → carbon dioxide + water

$$CH_4(g) + 2O_2(g) \rightarrow CO_2(g) + 2H_2O(g)$$

The gaseous alkanes are some of the most useful fuels. As you saw in Chapter 6, they are obtained by the fractional distillation of petroleum. Methane, better known as natural gas, is used for cooking as well as for heating offices, schools and homes (Figure 12.5a). Propane and butane burn with very hot flames and they are sold as liquefied petroleum gas (LPG). In rural areas where there is no supply of natural gas, central heating systems can be run on propane gas (Figure 12.5b). Butane, sometimes mixed with propane, is used in portable blowlamps and in gas lighters.

As stated, alkanes are generally unreactive but, like all homologous series they will undergo similar chemical reactions. They will react with halogens such as chlorine. Chlorine is quite a reactive non-metal and will react with, for example, methane in the presence of sunlight or ultraviolet light. This is known as a photochemical reaction, with the UV light providing the activation energy (E_a) for the reaction (see Chapter 6, p. 89). The overall chemical equation for this process is:

a   This is burning methane

b   Central heating systems can be run on propane

▲ **Figure 12.5**

methane + chlorine → chloromethane + hydrogen chloride

$$CH_4(g) + Cl_2(g) \rightarrow CH_3Cl(g) + HCl(g)$$

We can see that one hydrogen atom of the methane molecule is substituted (replaced) by a chlorine atom to form chloromethane (see Figure 12.6). This type of reaction is known as a **substitution reaction**. In a substitution reaction one atom or group of atoms is replaced by another atom or group of atoms: this is known as monosubstitution.

$$H - \overset{\displaystyle H}{\underset{\displaystyle H}{C}} - Cl$$

▲ **Figure 12.6** Chloromethane

Going further

Chloromethane is used extensively in the chemical industry. For example, chloromethane is used to make silicones. Silicones are used in building and construction. They are able to bond and seal materials such as concrete, glass, granite, plastics and steel (see Figure 12.7). This enables them to work better and last longer. There are dangers associated with substances containing chloromethane. It is a harmful substance and should be treated with caution.

▲ **Figure 12.7** Silicones are used carefully as adhesives as well as sealants

Early anaesthetics relied upon trichloromethane, $CHCl_3$, or chloroform. Unfortunately, this anaesthetic had a severe problem since the lethal dose was only slightly higher than that required to anaesthetise the patient. In 1956, halothane was discovered by chemists working at ICI. This is a compound containing chlorine, bromine and fluorine which has been used as an anaesthetic in recent years. Its formula is $CF_3CHBrCl$. However, even this is not the perfect anaesthetic since evidence suggests that prolonged exposure to this substance may cause liver damage. The search continues for even better anaesthetics.

A group of compounds called the chlorofluorocarbons (CFCs) were discovered in the 1930s. Because of their inertness they had many uses, especially as a propellant in aerosol cans. CFC-12 or dichlorodifluoromethane, CF_2Cl_2, was one of the most popular CFCs in use in aerosols. Scientists believe that CFCs released from aerosols are destroying the ozone layer and steps are being taken to reduce this threat.

The ozone hole problem

Our atmosphere protects us from harmful ultraviolet radiation from the Sun. This damaging radiation is absorbed by the relatively thin ozone layer found in the stratosphere (Figure 12.8b).

In the 1980s, large areas of low ozone concentrations were discovered in the atmosphere over Antarctica, Australasia and Europe. Scientists think that these holes in the ozone layer were produced by CFCs such as CFC-12. CFCs escape into the atmosphere and, because of their inertness, remain without further reaction until they reach the stratosphere and the ozone layer. In the stratosphere the high-energy ultraviolet radiation causes a chlorine atom to split off from the CFC molecule. This chlorine atom, or free radical, then reacts with the ozone.

$$Cl(g) + O_3(g) \rightarrow OCl(g) + O_2(g)$$

This is not the only problem with CFCs. They are also significant greenhouse gases (Chapter 11, p. 176). The ozone depletion and greenhouse effects have become such serious problems that an international agreement known as the Montreal Protocol on Substances that Deplete the Ozone Layer was agreed in 1987 to reduce the production and use of CFCs.

Research into replacements has taken place since then, producing better alternatives called hydrochlorofluorocarbons (HCFCs). These have lower ozone-depletion effects and are not effective greenhouse gases. The protocols of the 1980s have been superseded by the Kyoto Protocol of 1997 and the 2016 Paris Agreement, which has been signed by 197 countries. The holes in the ozone layer have recently started to recover (Figure 12.8a). It is believed that if more and more countries ban the use of CFCs then the ozone layer will recover by 2050.

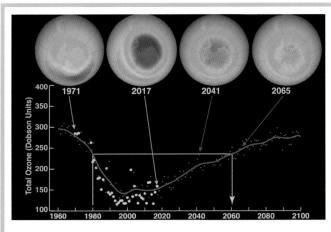

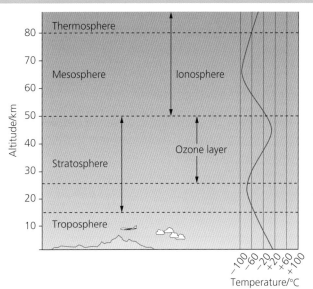

a This diagram comes from NASA's ozone monitoring programme, and simulations from a computer model. Note: Dobson units are a measure of the total amount of ozone in a vertical column from the ground to the top of the atmosphere

▲ **Figure 12.8**

b The ozone layer is between 25 km and 50 km above sea level

Other uses of alkanes

Besides their major use as fuels (p. 86), some of the heavier alkanes are used as waxes in candles, as lubricating oils and in the manufacture of another family of hydrocarbons – the alkenes.

> **Test yourself**
>
> **4** Write a balanced chemical equation to represent the combustion of propane.
> **5** In what mole proportions should chlorine and methane be mixed to produce chloromethane?
> **6** Draw the covalent bonding diagram for chloromethane, CH_3Cl.

Methane – another greenhouse gas

Methane, the first member of the alkanes, occurs naturally. Cows produce it in huge quantities when digesting their food. It is also formed by growing rice. Like carbon dioxide, it is a greenhouse gas (Chapter 11, p. 176) because it acts like the glass in a greenhouse (glasshouse) – it will let in heat from the Sun but will not let all of the heat back out again. It is thought that the greenhouse effect may contribute to global warming, which could have disastrous effects for life on this planet. A great debate is going on at the moment as to how we can reduce the amount of methane released into the atmosphere and hence reduce the effects of global warming.

> **Test yourself**
>
> **7** Use your research skills and your textbook to find out:
> **a** any other sources of methane found in nature
> **b** how climate change might affect your particular environment.

12.3 Alkenes

Alkenes form another homologous series of hydrocarbons of the general formula C_nH_{2n} where n is the number of carbon atoms. The alkenes are more reactive than the alkanes because they each contain a double carbon–carbon covalent bond between the carbon atoms (Figure 12.9). Molecules that possess a carbon–carbon double covalent bond, or even a triple covalent bond, are said to be **unsaturated**, because it is possible to break one of the two bonds to add extra atoms to the molecule. This feature of alkenes is very important and is responsible for the characteristic properties of these organic compounds. This feature is, for the alkenes, known as their functional group.

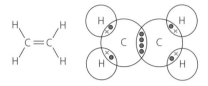

▲ **Figure 12.9** The bonding in ethene, the simplest alkene

The chemical test to show the difference between saturated and unsaturated hydrocarbons is discussed on p. 190.

> **Key definition**
>
> An **unsaturated** compound has molecules in which all carbon–carbon bonds are double bonds or triple bonds.

Naming the alkenes

All alkenes have names ending in -*ene*. Alkenes, especially ethene, are very important industrial chemicals. They are used extensively in the plastics industry and in the production of alcohols such as ethanol and propanol. See Table 12.2 and Figure 12.10.

▼ **Table 12.2** The first three alkenes and their physical properties

Alkene	Molecular formula	Melting point/°C	Boiling point/°C	Physical state at room temperature
Ethene	C_2H_4	–169	–104	Gas
Propene	C_3H_6	–185	–47	Gas
Butene	C_4H_8	–184	–6	Gas

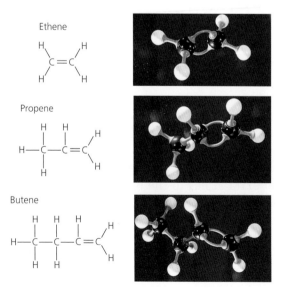

▲ **Figure 12.10** Displayed formula and shape of the first three alkenes

The structural formula of ethene, C_2H_4, is $CH_2=CH_2$ and the structural formula of propene is $CH_2=CHCH_3$.

Structural isomerism in alkenes

You saw when studying the alkanes that sometimes it is possible to write more than one displayed or structural formula to represent a molecular formula (p. 184). These are known as structural isomers. Butene and the higher alkenes also show structural isomerism. For example, there are three different compounds with the molecular formula C_4H_8. The displayed formulae of two of these substances along with their names and structural formulae are shown in Figure 12.11. The structural formulae of the isomers but-1-ene and but-2-ene give an unambiguous description of the way the atoms are arranged, including the location of the C=C double bond or functional group. The structural formulae are $CH_3CH_2CH=CH_2$ (but-1-ene) and $CH_3CH=CHCH_3$ (but-2-ene).

But-1-ene   But-2-ene

▲ **Figure 12.11** But-1-ene and but-2-ene: isomers of C_4H_8

> ### ➡ Going further
>
> The different isomers shown in Figure 12.11 have different melting and boiling points due to their different structures. All the alkene molecules with four or more carbon atoms possess isomers. As you can see from the structural formula of ethene and propene, it is not possible to produce isomers of these alkenes.

> ### ▶ Test yourself
>
> 8 Draw the displayed formulae for the isomers of C_5H_{10}.

Where do we get alkenes from?

Very few alkenes are found in nature. Most of the alkenes used by the petrochemical industry are obtained by breaking up larger, less useful alkane molecules obtained from the fractional distillation of petroleum. This is usually done by a process called **catalytic cracking**. In this process the alkane molecules to be 'cracked' (split up) are passed over a mixture of aluminium and chromium oxides heated to 550°C.

$$dodecane \longrightarrow decane + ethene$$

$$C_{12}H_{26}(g) \longrightarrow C_{10}H_{22}(g) + C_2H_4(g)$$

(found in kerosene)   shorter alkane alkene

Another possibility is:

$$C_{12}H_{26}(g) \longrightarrow C_8H_{18}(g) + C_4H_8(g)$$

In these reactions, hydrogen may also be formed during cracking. The amount of hydrogen produced depends on the conditions used. Since smaller hydrocarbons are generally in greater demand than the larger ones, cracking is used to match demand (Table 12.3).

▼ **Table 12.3** Percentages of the fractions in petroleum and the demand for them

Fraction	Approx % in petroleum	Approx % demand
Refinery gas	2	5
Gasoline	21	28
Kerosene	13	8
Diesel oil	17	25
Fuel oil and bitumen	47	34

This means that oil companies are not left with large surpluses of fractions containing the larger molecules.

Figure 12.12 shows the simple apparatus that can be used to carry out **thermal cracking** reactions in the laboratory. You will notice that in the laboratory we may use a catalyst of broken, unglazed pottery. This experiment should only ever be carried out as a teacher demonstration.

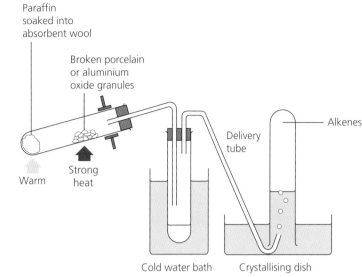

Paraffin soaked into absorbent wool
Broken porcelain or aluminium oxide granules
Warm
Strong heat
Delivery tube
Alkenes
Cold water bath
Crystallising dish

▲ **Figure 12.12** The cracking of an alkane in the laboratory

Test yourself

9 Using the information in Table 12.2 (p. 188), estimate the boiling point of pentene.
10 Write a balanced chemical equation to represent the process that takes place when decane is cracked.

12.4 The chemical behaviour of alkenes

The double bond makes alkenes more reactive than alkanes in chemical reactions. For example, hydrogen *adds* across the double bond of ethene, under suitable conditions, forming ethane (Figure 12.13). This type of reaction is called an **addition reaction**. The usefulness of this type of reaction is that only one product is formed.

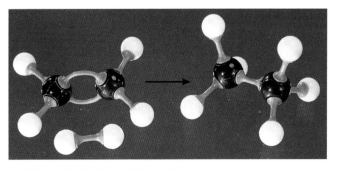

▲ **Figure 12.13** The addition of hydrogen to ethene using molecular models

Addition reactions

Hydrogenation

Hydrogen adds across the C=C double bond in ethene. This reaction is called hydrogenation. The conditions necessary for this reaction to take place are a temperature of 200°C in the presence of a nickel catalyst.

$$\text{ethene} + \text{hydrogen} \longrightarrow \text{ethane}$$

$$C_2H_4(g) + H_2(g) \longrightarrow C_2H_6(g)$$

$$\begin{array}{c} H \\ \diagdown \\ C=C \\ \diagup \quad \diagdown \\ H \qquad H \end{array} + H-H \longrightarrow \begin{array}{c} H \ H \\ | \ | \\ H-C-C-H \\ | \ | \\ H \ H \end{array}$$

Hydrogenation reactions like the one shown with ethene are used in the manufacture of margarines from vegetable oils.

Hydration

Another important addition reaction is the one used in the manufacture of ethanol. Ethanol has important uses as a solvent and a fuel. It is formed when water (as steam) is added across the double bond in ethene in the presence of an acid catalyst. For this reaction to take place, the reactants have to be passed over a catalyst of phosphoric(V) acid (absorbed on silica pellets) at a temperature of 300°C and pressure of 6000 kPa.

$$\text{ethene} + \text{steam} \xrightleftharpoons[\substack{\text{phosphoric(v)} \\ \text{acid catalyst}}]{\text{300°C, 6000 kPa}} \text{ethanol}$$

$$C_2H_4(g) + H_2O(g) \rightleftharpoons C_2H_5OH(g)$$

$$\begin{array}{c} H \\ \diagdown \\ C=C \\ \diagup \quad \diagdown \\ H \qquad H \end{array} + H-OH \rightleftharpoons \begin{array}{c} H \ H \\ | \ | \\ H-C-C-H \\ | \ | \\ H \ OH \end{array}$$

This reaction is reversible as is shown by the equilibrium (⇌) sign. The conditions have been chosen to ensure the highest possible yield of ethanol. In other words, the conditions have been chosen so that they favour the forward reaction. The percentage yield is high at approximately 96%. It should be noted, however, that the ethene is a non-renewable resource.

For a further discussion of ethanol and alcohols generally, see Chapter 13, p. 199.

This method of manufacturing ethanol is a continuous process. In a continuous process, reactants are continually fed into the reaction vessel or reactor as the products are removed. Generally, continuous processing is employed if the substance being made is needed on a large scale.

As you have seen, during addition reactions only one product is formed.

Halogenation – a test for unsaturated compounds

The addition reaction between bromine dissolved in an organic solvent, or water, and alkenes is used as a chemical test for the presence of a double bond between two carbon atoms. When a few drops of this bromine solution are shaken with the hydrocarbon, if it is an alkene (such as ethene) a reaction takes place in which bromine joins to the alkene double bond. This results in the bromine solution losing its red/brown colour. If an alkane, such as hexane, is shaken with a bromine solution of this type, no colour change takes place (Figure 12.14). This is because there are no double bonds between the carbon atoms of alkanes.

$$\text{ethene} + \text{bromine} \longrightarrow \text{dibromoethane}$$

$$C_2H_4(g) + Br_2 \text{ (in solution)} \longrightarrow C_2H_4Br_2 \text{ (in solution}$$

$$\begin{array}{c} H \qquad H \\ \diagdown \quad \diagup \\ C=C \\ \diagup \quad \diagdown \\ H \qquad H \end{array} + Br-Br \longrightarrow \begin{array}{c} H \ H \\ | \ | \\ H-C-C-H \\ | \ | \\ Br \ Br \end{array}$$

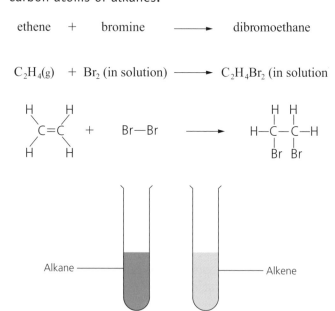

▲ **Figure 12.14** The alkane, has no effect on the bromine solution, but the alkene decolourises it

Test yourself

11 What is meant by the term 'addition reaction'?
12 How would you test the difference between ethane and ethene?
13 Write the displayed formula for pentene.
14 Which of the following organic chemicals are alkanes and which are alkenes?
   – Propene, C_3H_6 – Nonane, C_9H_{20}
   – Octane, C_8H_{18} – Butene, C_4H_8

   State why you have chosen your answers.
15 The following displayed formulae are those of some organic molecules you have seen in this chapter.

(a)

H—C—C—OH with H H above top carbons and H H below

(b)

C=C with H H on left and H H on right

(c)

H—C—C—H with H H above and H H below

Identify by name each of the molecules and write their molecular formula.

Practical skills

Alkanes and alkenes

Safety

- Eye protection must be worn.

A student wrote the following account of an experiment carried out by the teacher to test which of two liquid hydrocarbons was an alkane and which was an alkene.

To a few drops of each liquid in separate test tubes, they added <u>a few drops</u> of <u>red aqueous bromine solution</u> from a measuring cylinder. They then corked the test tubes and shook them <u>a little</u>. In one test tube the <u>solution</u> went colourless showing this hydrocarbon to be an <u>alkane</u> while there was no change in colour of the other, so this must be the <u>alkene</u>.

Some statements in the student's account have been underlined by the teacher.

Answer these questions to identify how the experiment could be improved.

1 Which major safety precaution did they forget to mention at the start of the experiment?
2 Do you think the teacher would have added a few drops from a measuring cylinder? Explain your answer.
3 What would be the actual colour of the aqueous bromine solution?
4 How would the teacher have shaken the test tubes?
5 Would you expect only one layer in the solution after adding the aqueous bromine solution? Explain your answer.
6 Has the student got the identification of the alkane and the alkene correct? Explain your answer.

12.5 Polymers

Polythene is a **plastic** that was discovered by accident. Through the careful examination of this substance, when it was accidentally discovered, the plastics industry was born. Polythene is now produced in millions of tonnes worldwide every year. It is made by heating ethene to a relatively high temperature under a high pressure in the presence of a catalyst.

   In a chemical equation to show this process, we show that a large number of ethene molecules are adding together by using the letter '*n*' to indicate very large numbers of monomer molecules.

$$n \left(\begin{array}{c} H \quad\quad H \\ C=C \\ H \quad\quad H \end{array} \right) \longrightarrow \left(\begin{array}{c} H\ H \\ -C-C- \\ H\ H \end{array} \right)_n$$

The polymer formed, in the chemical equation, is shown as a repeat unit which repeats *n* times to form the polymer chain.

When small molecules like ethene join together to form long chains of atoms, called **polymers**, the process is called **polymerisation**. The small molecules, like ethene, which join together in this way are called **monomers**. A polymer chain, a very large molecule or a macromolecule, often consists of many thousands of monomer units and in any piece of plastic there will be many millions of polymer chains. Since in this polymerisation process the monomer units add together to form only one product, the polymer, the process is called **addition polymerisation**.

> **Key definitions**
>
> **Polymers** are large molecules built up from many small units called monomers.
>
> **Plastics** are made from polymers.

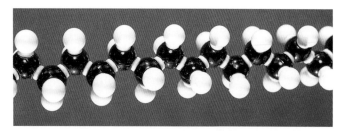

▲ **Figure 12.15** This model shows part of the poly(ethene) polymer chain

Poly(ethene), like many other addition polymers has many useful properties being tough, easy to mould and an excellent insulator. One of the drawbacks however, as you will see later in this section, is that it is not affected by the weather and does not corrode, and therefore has disposal problems.

It can be found as a substitute for natural materials in plastic bags, sandwich boxes, washing-up bowls, wrapping film, milk-bottle crates and washing-up liquid bottles (Figure 12.16).

▲ **Figure 12.16** These crates are made from poly(ethene)

Other alkene molecules can also produce substances like poly(ethene); for example, propene produces poly(propene), which is used to make ropes and packaging.

In theory, any molecule that contains a carbon–carbon double covalent bond can form an addition polymer. For example, ethene, $CH=CH_2$, will undergo an addition reaction to form poly(ethene).

The actual structure of the polymer can only be shown/represented by drawing the monomer as shown in Figure 12.17. In this figure you will see that the double bond has been replaced by a single bond and there are two extension bonds drawn on either side, which show that the polymer chain extends in both those directions.

Ethene
(monomer)

Repeat unit in
the polymer

▲ **Figure 12.17** Addition polymerisation of ethene

 Going further

Other addition polymers

Many other addition polymers have been produced. Often the plastics are produced with particular properties in mind, such as PVC (polyvinyl chloride or poly(chloroethene) and PTFE (poly(tetrafluoroethene)). Both of these plastics have monomer units similar to ethene.

PVC monomer
(vinyl chloride or
chloroethene)

PTFE monomer
(tetrafluoroethene)

If we use chloroethene (Figure 12.18a), the polymer we make is slightly stronger and harder than poly(ethene) and is particularly good for making pipes for plumbing (Figure 12.19).

a Model of chloroethene, the PVC monomer

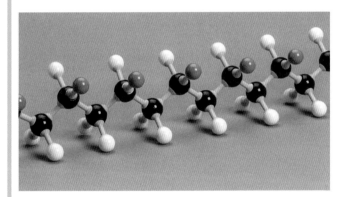

b Model of part of a PVC polymer chain

▲ **Figure 12.18**

▲ **Figure 12.19** These pipes are made from PVC

PVC is the most versatile plastic and is the second most widely used, after poly(ethene). Worldwide more than 27 million tonnes are produced annually.

If we start from tetrafluoroethene (Figure 12.20a), the polymer we make, PTFE, has some slightly unusual properties:

- it will withstand very high temperatures, of up to 260°C
- it forms a very slippery surface
- it is hydrophobic (water repellent)
- it is highly resistant to chemical attack.

These properties make PTFE an ideal 'non-stick' coating for frying pans and saucepans. Every year more than 150 000 tonnes of PTFE are made.

$$n \left(\begin{array}{c} F \quad\quad F \\ \diagdown \quad \diagup \\ C = C \\ \diagup \quad \diagdown \\ F \quad\quad F \end{array} \right) \longrightarrow \left(\begin{array}{c} F \quad F \\ | \quad\;\; | \\ -C-C- \\ | \quad\;\; | \\ F \quad F \end{array} \right)_n$$

Monomer   Polymer chain

The properties of some addition polymers along with their uses are given in Table 12.4.

a Model of tetrafluoroethene, the PTFE monomer

▲ **Figure 12.20**

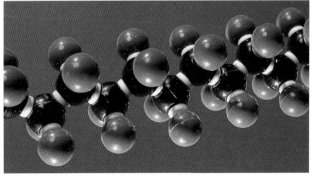

b Model of part of a PTFE polymer chain

▼ **Table 12.4** Some addition polymers

Plastic	Monomer	Properties	Uses
Poly(ethene)	$CH_2=CH_2$	Tough, durable	Carrier bags, bowls buckets, packaging
Poly(propene)	$CH_3CH=CH_2$	Tough, durable	Ropes, packaging
PVC	$CH_2=CHCl$	Strong, hard (less flexible than poly(ethene))	Pipes, electrical insulation, guttering
PTFE	$CF_2=CF_2$	Non-stick surface, withstands high temperatures	Non-stick frying pans, soles of irons
Polystyrene	$CH_2=CHC_6H_5$	Light, poor conductor of heat	Insulation, packaging (especially as foam)
Perspex	$CH_2=C(CO_2CH_3)CH_3$	Transparent	Used as a glass substitute

Environmental challenges of plastics

In the last 30 to 40 years, plastics have taken over as replacement materials for metals, glass, paper and wood, as well as for natural fibres such as cotton and wool. This is not surprising since plastics are light, cheap, relatively unreactive, can be easily moulded and can be dyed bright colours. Unfortunately, as you saw earlier in this section, they are not affected by the weather and so do not corrode. This is both an advantage, in the way they can be used, and a disadvantage, as it makes plastics so difficult to dispose of safely. Plastics have contributed significantly to the household waste problem, at least 10% in some countries, and it is getting worse (Figure 12.21)!

▲ **Figure 12.21** This plastic waste is ready to go to landfill

In the recent past, much of our plastic waste has been used as landfill in, for example, disused quarries. However, all over the world these sites are getting very much harder to find and it is becoming more and more expensive to dispose of the waste. Also the older landfill sites are now beginning to produce gases, including methane (which is a greenhouse gas), which contribute to global warming. The alternatives to dumping plastic waste are certainly more economical and more satisfactory but also create their problems.

›› Incineration schemes have been developed to use the heat generated from burning waste for heating purposes (Figure 12.22). However, problems with the combustion process (which can result in the production of toxic gases) mean that especially high temperatures have to be employed during the incineration process driving energy costs up.

›› Recycling produces large quantities of black plastic bags and sheeting for resale. However there are problems with recycling in that some plastics cannot be recycled because of their properties.

›› **Biodegradable** plastics, as well as those polymers that degrade in sunlight (photodegradable, Figure 12.23a), have been developed. Other common categories of degradable plastics include synthetic biodegradable plastics which are broken down by bacteria, as well as plastics which dissolve in water (Figure 12.23b). The property that allows plastic to dissolve in water has been used in relatively new products, including soluble capsules containing liquid detergent. Very recently chemists have found ways of involving carbon dioxide and sugars in the production of plastics that are also broken down by bacteria. However, the vast majority of polymers are still non-biodegradable.

There is a further huge global problem which has been highlighted recently. That is, there are extreme pollution problems caused by non-biodegradable plastics and their accumulation in the world's oceans (see Figure 12.24). Aquatic life is being decimated in some regions of the world. The problems with plastics are causing a dramatic rethink of the role of plastics in our society!

▲ **Figure 12.22** Incineration plants can burn waste material to produce heat which can be used to provide warm water or can be converted into electricity

a   This plastic bag is photodegradable

▲ **Figure 12.23**

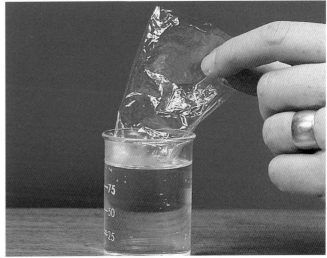

b   This plastic dissolves in water

▲ **Figure 12.24** Non-biodegradable plastics accumulate in the oceans of the world

> ## Test yourself
>
> 16 Draw the structure of the repeat unit of the addition polymer formed from $CH_3—CH_2CH=CH_2$.
> 17 Draw the displayed formula of the monomer from which the addition polymer below has been produced.

$$
\begin{array}{cccccc}
C_6H_5 & H & C_6H_5 & H & C_6H_5 & H \\
| & | & | & | & | & | \\
—C— & C— & C— & C— & C— & C— \\
| & | & | & | & | & | \\
H & H & H & H & H & H
\end{array}
$$

Revision checklist

After studying Chapter 12 you should be able to:

✔ Draw and interpret the displayed formula of a molecule to show all of the atoms and all of the bonds.

✔ Write and interpret general formulae to show the ratio of atoms in the molecules of the compounds in a homologous series.

✔ Identify a functional group as an atom or group of atoms that determine the chemical properties of a homologous series.

✔ State that a structural formula is a description of the way the atoms in a molecule are arranged, including $CH_2=CH_2$.

✔ Define structural isomers as compounds with the same molecular formula, but different structural formulae, including C_4H_{10} as $CH_3CH_2CH_2CH_3$ and $CH_3CH(CH_3)CH_3$ and C_4H_8 as $CH_3CH_2CH=CH_2$ and $CH_3CH=CHCH_3$.

✔ Describe a homologous series as compounds that have the same functional group, the same general formula, differ from one member to the next by a –CH_2– unit , show a trend in physical properties and have similar chemical properties.

✔ State that a saturated compound has molecules in which all carbon–carbon bonds are single bonds.

✔ State that an unsaturated compound has molecules in which one or more carbon–carbon bonds are double bonds or triple bonds.

✔ Name and draw the structural and displayed formulae of unbranched alkanes and alkenes and the products of their reactions containing up to four carbon atoms per molecule.

✔ State the type of compound present given the chemical name ending in -ane, -ene, or from a molecular, structural or displayed formula.

✔ State that the bonding in alkanes is single covalent and that alkanes are saturated hydrocarbons.

✔ Describe the properties of alkanes as being generally unreactive, except in terms of combustion and substitution by chlorine.

✔ State that in a substitution reaction one atom or group is replaced by another atom or group.

✔ Describe the substitution reaction of alkanes with chlorine as a photochemical reaction, with UV light providing the activation energy, and draw the structural or displayed formulae of the products, limited to monosubstitution.

✔ State that the bonding in alkenes includes a double carbon–carbon covalent bond and that alkenes are unsaturated hydrocarbons.

✔ Describe the manufacture of alkenes and hydrogen by the cracking of larger alkanes using a high temperature, 550°C, and a catalyst.

✔ Describe the reasons for the cracking of larger alkanes.

✔ Distinguish between saturated and unsaturated hydrocarbons using aqueous bromine.

✔ State that in an addition reaction, only one product is formed.

✔ Describe the properties of alkenes in terms of addition reactions with bromine, hydrogen and steam, and draw the structural or displayed formulae of the products.

✔ Define polymers as large molecules built up from many small units called monomers.

✔ Describe the formation of poly(ethene) as an example of addition polymerisation.

✔ Identify the repeat units and/or linkages in polymers.

✔ Deduce the structure or repeat unit of an addition polymer from a given alkene and vice versa.

✔ State that plastics are made from polymers.

✔ Describe how the properties of plastics have implications for their disposal.

✔ Describe the environmental challenges caused by plastics in terms of their disposal in landfill, accumulation in oceans and formation of toxic gases from burning.

Exam-style questions

1 Explain the following statements:
   a Ethene is called an unsaturated
      hydrocarbon. [2]
   b The cracking of larger alkanes into simple
      alkanes and alkenes is important to the
      petrochemical industry. [3]
   c The conversion of ethene to ethanol is an
      example of an addition reaction. [2]

2 The following question is about some of the
   reactions of ethene.

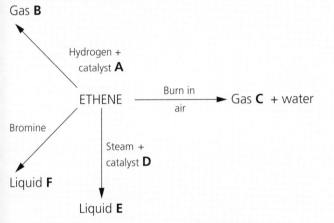

Gas **B**

Hydrogen +
catalyst **A**

ETHENE ——Burn in air——> Gas **C** + water

Bromine

Steam +
catalyst **D**

Liquid **F**

Liquid **E**

   a Give the names and formulae for
      substances **A** to **F**. [12]
   b i Give a word and balanced chemical
         equation to represent the reaction in
         which liquid **E** is formed. [4]
     ii Identify the reaction conditions
         required for the process to take place. [3]
   c Give the homologous series that gas **B**
      belongs to. [1]
   d Describe a chemical test which would
      allow you to identify gas **C**. [2]

3 a Petroleum is a mixture of *hydrocarbons* which
      belong to the *homologous series* called the
      alkanes.
      Explain the meaning of the terms in italics. [6]
   b Alkanes can be converted into substances
      which are used as solvents. To do this the
      alkane is reacted with a halogen, such as
      chlorine, in the presence of ultraviolet light.

   i Give a word and balanced chemical
      equation for the reaction between methane
      and chlorine. [4]
   ii Identify the type of reaction taking place. [1]

4 Ethene, C_2H_4, is the starting material for making
   plastic carrier bags.

$$n \left(\begin{array}{c} H \quad\quad H \\ \diagdown \quad\ \diagup \\ C=C \\ \diagup \quad\ \diagdown \\ H \quad\quad H \end{array} \right) \longrightarrow \left(\begin{array}{c} H \quad H \\ | \quad\ | \\ -C-C- \\ | \quad\ | \\ H \quad H \end{array} \right)_n$$

   a Identify the type of chemical change taking
      place in the diagram above. [1]
   b Identify the product formed by this
      reaction. [1]
   c i The alkene, ethene, is made by cracking
         large alkane molecules. Describe a simple
         chemical test to show that ethene is
         present. [2]
     ii Draw the displayed formula of any
         products formed in part **a**. [1]

5 a The majority of plastic carrier bags are
      difficult to dispose of.
      i Explain why carrier bags should not just
         be thrown away. [2]
     ii Explain why plastic carrier bags should
         be recycled. [2]
    iii Give one advantage that a plastic carrier
         bag has over one made out of paper. [1]
   b A label like the one below is found on
      some plastic carrier bags.
      'This plastic carrier bag is made from a
      substance that is made from the chemical
      elements carbon and hydrogen only. When
      the carrier bag is burned it produces carbon
      dioxide and water. These substances are
      natural and will not harm the environment.'
      i Give the meaning of the term 'element'. [2]
     ii Identify the name given to the type of
         compound that contains the elements
         carbon and hydrogen only. [1]

iii When the plastic bag burns, heat energy is given out. Give the name used to describe reactions that give out heat energy. [1]

iv The plastic bag will probably give off a toxic gas when it is burned in a limited supply of air. Give the name and formula of this gas. [2]

6 a Identify which of the following formulae represent alkanes, which are alkenes, and which represent neither.

CH_3 $C_{12}H_{24}$

C_6H_{12} $C_{20}H_{42}$

C_5H_{12} C_2H_4

C_6H_6 C_8H_{18}

C_9H_{20} C_3H_7 [3]

b Draw the displayed formula for all the possible isomers which have the molecular formula C_6H_{14}. [10]

13 Organic chemistry 2

FOCUS POINTS

★ What are the structures and formulae of the homologous series, alcohols and carboxylic acids?
★ What are the uses of ethanol and how can it be manufactured?
★ How does ethanoic acid react with metals, bases and carbonates?
★ How do condensation polymers differ from addition polymers and why is this reaction important in the production of polyamides and polyesters?
★ Why can proteins be described as a type of polymer?

In Chapter 12 we discussed the organic compounds obtained from petroleum, the alkanes, and how they are converted into another homologous series of hydrocarbons called the alkenes. In this chapter you will learn that if you replace one of the hydrogen atoms on an alkane molecule with a group such as –OH, the hydroxyl group, then you get a new homologous series called the alcohols. Also if you replace one of the hydrogen atoms on an alkane molecule with a –COOH group then you get a homologous series called the carboxylic acids. You will see that whichever group you have attached, it will bring with it a new set of physical and chemical properties. These groups are known as functional groups and it is this group of atoms that is responsible for the characteristic reactions of the organic compound. You will also study the formation and uses of another type of polymer known as condensation polymers. Finally, you will learn about the natural condensation polymers, proteins.

13.1 Functional groups

You saw in Chapter 12 that the functional group of an organic molecule is an atom or group of atoms that determine the chemical properties of a homologous series. It is not surprising, therefore, that there are so many organic molecules given that there are many functional groups that can replace an H atom of basic hydrocarbon molecules.

Table 13.1 shows some examples of functional groups. In the table, R represents an alkyl group or a hydrogen atom. An alkyl group has the general formula based on the alkanes, C_nH_{2n+1}. When $n = 1$, $R = CH_3$; when $n = 2$, $R = C_2H_5$, and so on.

▲ **Figure 13.1** This fruit juice contains plenty of vitamin C, or ascorbic acid, which contains the functional group –COOH

▼ **Table 13.1** The functional groups present in some homologous series of organic compounds

Class of compound	Functional group
Alcohols	R–OH
Carboxylic acids	R–COOH
Esters	R–COOR
Amines	R–NH$_2$
Amides	R–CONH$_2$

13.2 Alcohols (R–OH)

The alcohols form another homologous series with the general formula $C_nH_{2n+1}OH$ (or R–OH, where R represents an alkyl group). All the alcohols possess an –OH as the functional group. Table 13.2 shows the names and condensed formulae of the first four members along with their melting and boiling points.

▼ **Table 13.2** Some members of the alcohol family

Alcohol	Formula	Melting point/°C	Boiling point/°C
Methanol	CH_3OH	−94	64
Ethanol	CH_3CH_2OH	−117	78
Propanol	$CH_3CH_2CH_2OH$	−126	97
Butanol	$CH_3CH_2CH_2CH_2OH$	−89	117

Test yourself

1 Which of the following are alcohols?
 pentanol, cyclohexane, butene, heptanol
2 What is the name of the alcohol with the molecular formula $C_8H_{17}OH$?
3 Looking at the molecular formulae of the following compounds, which is an alcohol?
 C_5H_{12} C_2H_5OH C_4H_{10} C_7H_{14}

The alcohols are named by reference to the corresponding alkane (see Chapter 12, p. 184), always ensuring that the hydrocarbon chain is numbered from the end that gives the lowest number to the position of the −OH group. If the −OH group is positioned at the end of the alcohol, for example in $CH_3CH_2CH_2OH$, then the position of this group is shown by numbering the carbon atom it is attached to as '1'. This molecule is called propan-1-ol, or propanol for short. If the −OH group is attached to the second carbon atom, $CH_3CH(OH)$ CH_3, then this molecule is called propan-2-ol. The displayed formulae of these two isomers of propanol are shown in Figure 13.2 along with those for two isomers of butanol.

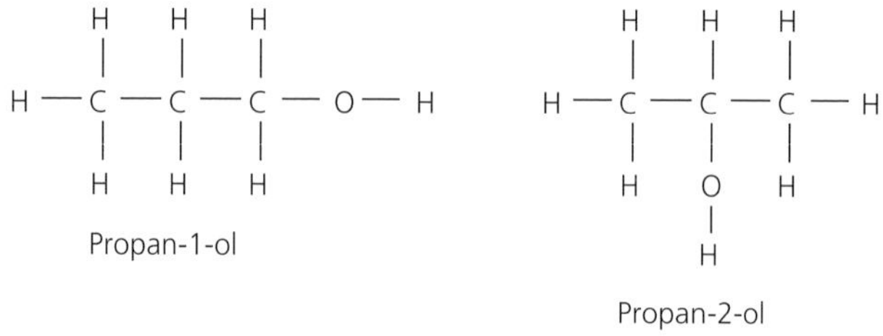

a Propan-1-ol and propan-2-ol

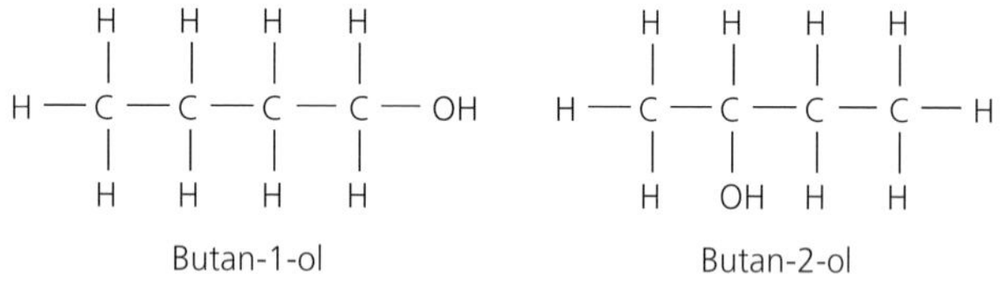

b Butanol-1-ol and butan-2-ol

▲ **Figure 13.2** Displayed formulae of alcohols

Alcohols have high boiling points and relatively low volatility. Alcohol molecules are like water molecules (H–OH) in that they are polar (see Chapter 11, p. 166).

Going further

Alcohol molecules are polar because of the presence of the −OH group, in which the hydrogen attached to oxygen creates (within this group) a small difference in charge (Figure 13.3). Other organic molecules that are polar are carboxylic acids such as ethanoic acid (p. 204).

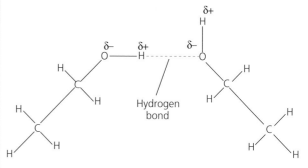

▲ **Figure 13.3** Polar −OH groups in ethanol molecules lead to hydrogen bonding and an attraction between neighbouring molecules

Because of the presence of the polar −OH groups, there is a relatively strong polar attraction between alcohol molecules. This polar attraction between the charges δ+ and δ− in neighbouring molecules is called a hydrogen bond. This means that the molecules have a much stronger attraction for each other than molecules of the corresponding alkane of similar relative molecular mass, M_r. For example, ethanol has an M_r value of 46 and is a liquid at room temperature with a boiling point of 78°C, while propane ($CH_3CH_2CH_3$) has an M_r of 44 and is a gas at room temperature with a boiling point of −42°C.

It is possible to think of alcohol molecules as water molecules in which an H atom has been replaced by an alkyl group, for example $–C_2H_5$ in C_2H_5OH. This close similarity between their molecules explains why water and alcohols with small molecules, such as methanol and ethanol, are miscible. They mix because of the presence of the polar –OH group in both molecules.

Many other materials, such as food flavourings, are made from ethanol. As ethanol is also a very good solvent and evaporates easily, it is used extensively as a solvent for paints, glues, aftershave and many other everyday products (Figure 13.4).

▲ **Figure 13.4** This aftershave contains alcohol

Ethanol is by far the most important of the alcohols and is often just called 'alcohol'. Ethanol can be produced by fermentation (p. 202) as well as by the hydration of ethene (p. 203). It is a neutral, colourless, volatile liquid which does not conduct electricity.

Combustion

The combustion of ethanol is an important property of ethanol. Ethanol burns quite readily with a clean, hot flame.

$$\text{ethanol} \quad + \text{oxygen} \rightarrow \quad \text{carbon} + \quad \text{water} \ + \text{energy}$$
$$\text{dioxide}$$

$$CH_3CH_2OH(l) + 3O_2(g) \rightarrow 2CO_2(g) + 3H_2O(g) + \text{energy}$$

As methylated spirit, it is used in domestic burners for heating and cooking. Methylated spirit is ethanol with small amounts of poisonous substances, such as methanol, added to it. Some countries, like Brazil and the US, already use ethanol mixed with petrol as a fuel for cars and this use is increasing worldwide.

Oxidation

Ethanol, if left open to the atmosphere, oxidises. Vinegar is manufactured by allowing solutions containing alcohol to oxidise. Bacteria present in the solution aid this process to take place. This type of oxidation is how different vinegars are produced.

Ethanol can also be oxidised to ethanoic acid (an organic acid also called acetic acid) by powerful oxidising agents, such as acidified potassium manganate(VII). During the reaction the purple colour of potassium manganate(VII)) is removed (Figure 13.5) as the ethanol is oxidised to ethanoic acid.

$$\text{ethanol} \quad + \text{oxygen} \xrightarrow{\text{heat}} \quad \text{ethanoic} \ + \text{water}$$
$$\text{(from} \qquad\qquad \text{acid}$$
$$\text{potassium}$$
$$\text{manganate}$$
$$\text{(VII))}$$

$$CH_3CH_2OH(l) \quad + \quad 2[O] \longrightarrow CH_3COOH(aq) + H_2O(l)$$

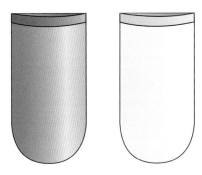

Carboxylic acid group

▲ **Figure 13.5** Potassium manganate(VII) turns from purple to colourless

> ## Test yourself
>
> 4 Why does ethanol make such a good solvent?

Cholesterol – a complex molecule that contains the –OH group

Cholesterol is a naturally occurring and essential chemical. It belongs to a family of chemicals called steroids and also contains an alcohol group (Figure 13.6). Cholesterol is found in almost all of the tissues in the body, including nerve cells. Levels of cholesterol above normal (above 6.5 mmol/l) are associated with an increased risk of heart disease. Cholesterol hardens and blocks off arteries by building up layers of solid material (atheroma) inside the arteries (Figure 13.7). This is particularly serious if the arteries that supply the heart or brain are blocked. Simple tests are now available to monitor cholesterol levels. People with high levels can be treated and can follow special low-fat and low-cholesterol diets.

Cholesterol

▲ **Figure 13.6** The displayed formula of cholesterol

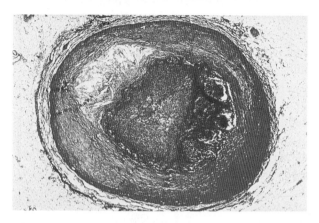

a This artery is being blocked by atheroma, which may be related to high levels of cholesterol in the blood

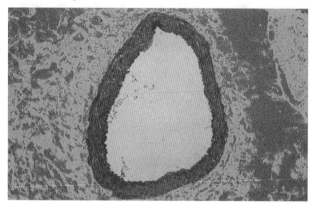

b This is a healthy artery

▲ **Figure 13.7**

Manufacture of ethanol

Biotechnology involves making use of micro-organisms or their components, such as enzymes, for the benefit of humans to produce, for example, foods such as yoghurt and bread. One of the oldest biotechnologies is that of **fermentation**. It involves a series of biochemical reactions catalysed by enzymes in yeast, a micro-organism.

Fermentation

Fermentation in the laboratory can be carried out using sugar solution. Yeast is added to the solution. The yeast uses the sugar for energy during **anaerobic respiration** (respiration without oxygen), and so the sugar is broken down to give carbon dioxide and ethanol. The best temperature for this

process to be carried out is 25–35°C. This is because the enzymes in yeast which catalyse this process will be denatured (destroyed) at temperatures much above this, for example, at 40°C.

$$\text{glucose} \xrightarrow{\text{yeast}} \text{ethanol} + \text{carbon dioxide}$$

$$C_6H_{12}O_6(aq) \longrightarrow 2C_2H_5OH(l) + 2CO_2(g)$$

The process in industry by which glucose is fermented is a **batch process**. A batch process is one in which each batch goes through one stage of the production process before moving onto the next stage. So in this case, ethanol is the primary product. After fermentation of the glucose has finished, the ethanol in this batch has to be separated from the mixture by fractional distillation. The process then is repeated with another batch of reagents. Batch processes are used where small amounts of a substance are required. The percentage yield by this process is approximately 15%. Figure 13.8 shows a simple apparatus for obtaining ethanol from glucose in the laboratory.

▲ **Figure 13.8** Fermenting glucose and yeast to produce ethanol. The bag is inflated during the experiment by CO_2

Hydration of ethene

Another important addition reaction is the hydration of ethene used in the manufacture of ethanol. Ethanol has important uses as a solvent and a fuel. It is formed when water (as steam) is added across the double bond in ethene in the presence of an acid catalyst. For this reaction to take place, the reactants have to be passed over an acid catalyst (usually phosphoric(V) acid absorbed on silica pellets) at a temperature of 300°C and pressure of 6000 kPa.

$$\text{ethene} \quad + \quad \text{steam} \quad \underset{\substack{\text{phosphoric(V)}\\ \text{acid catalyst}}}{\overset{300°C,\ 6000\ kPa}{\rightleftharpoons}} \quad \text{ethanol}$$

$$C_2H_4(g) \quad + \quad H_2O(g) \quad \rightleftharpoons \quad C_2H_5OH(g)$$

For a further discussion of this reaction, see Chapter 12, p. 190.

As discussed earlier, ethanol is a very useful chemical and used in many ways. It can be manufactured by two processes: fermentation of glucose and hydration of ethene. A summary of the methods of manufacture are shown in Table 13.3.

▼ **Table 13.3** Summary of the two methods that are used to make ethanol

	Fermentation	Hydration
Conditions employed	37°C and so less energy needed compared to the hydration method	300°C, 6000 kPa uses a catalyst of phosphoric acid
Processing	Manufactured in batches	Manufactured by a continuous process
Sustainability	Sustainable source – the glucose is renewable	Finite source – ethene is non-renewable
Purification	Fractional distillation	Becomes pure during production
Percentage yield	Low – about 15%	High – about 96%

As you would expect, both methods of preparation of ethanol have their advantages and disadvantages. Table 13.4 shows some of these.

▼ **Table 13.4** Advantages and disadvantages of preparation methods for ethanol

	Advantages	Disadvantages
Fermentation	• Low energy consumption • Uses readily available materials	• Slow process • Not a continuous process
Hydration of ethene	• Much faster process • Continuous process	• High energy consumption • Requires ethene which has to be obtained by cracking

> ## Test yourself
>
> 5 Why should being a continuous process be an advantage for a process for making ethanol?

Going further

Baking

To make bread, fresh yeast is mixed with warm sugar solution and the mixture is added to the flour. This dough mixture is then put into a warm place to rise. The dough rises due to the production of carbon dioxide from aerobic respiration (respiration with oxygen) by the yeast. The products of this style of respiration are different to those of anaerobic respiration.

$$\text{sugar} + \text{oxygen} \xrightarrow{\text{yeast}} \text{carbon dioxide} + \text{water} + \text{energy}$$

$$C_6H_{12}O_6(aq) + 6O_2(g) \longrightarrow 6CO_2(g) + 6H_2O(l) + \text{energy}$$

After the dough has risen, it is baked: the heat kills the yeast and the bread stops rising.

New applications of biotechnology

A large number of firms throughout the world are investing large sums of money in the newer biotechnology applications now in use.

» Enzymes can be isolated from micro-organisms and used to catalyse reactions in other processes. For example, proteases are used in biological detergents to digest protein stains such as blood and food. Also, catalase is used in the rubber industry to help convert latex into foam rubber.

» Our ability to manipulate an organism's genes to make it useful to us is called genetic engineering. This is being used, for example, to develop new varieties of plants for agriculture as well as making important human proteins such as insulin and growth hormone.

However, a word of caution is necessary. The new biotechnologies may not be without dangers. For example, new pathogens (organisms that cause disease) might be created accidentally. Also, new pathogens may be created deliberately for use in warfare. As you can imagine, there are very strict guidelines covering these new biotechnologies, especially in the area of research into genetic engineering.

13.3 Carboxylic acids

The **carboxylic acids** form another homologous series, this time with the general formula $C_nH_{2n+1}COOH$. All the carboxylic acids possess $-COOH$ as their functional group. Table 13.5 shows the first four members of this homologous series along with their melting and boiling points. Figure 13.9 shows the actual arrangement as displayed formulae of the atoms in these members of this family.

▼ **Table 13.5** Some members of the carboxylic acid series

Carboxylic acid	Structural formula	Melting point/ °C	Boiling point/ °C
Methanoic acid	HCOOH	9	101
Ethanoic acid	CH_3COOH	17	118
Propanoic acid	CH_3CH_2COOH	–21	141
Butanoic acid	$CH_3CH_2CH_2COOH$	–6	164

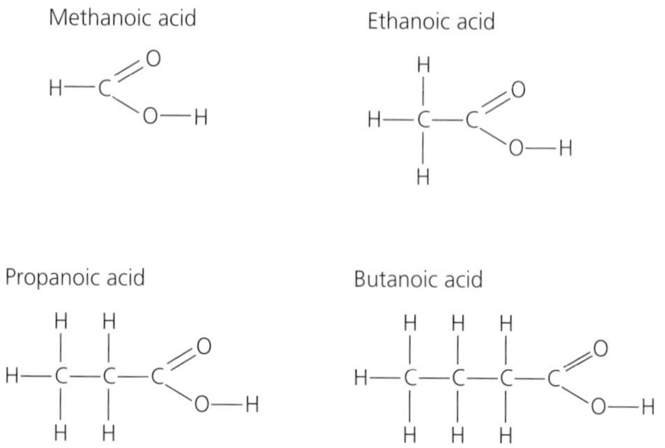

▲ **Figure 13.9** The displayed formulae of the first four members of the carboxylic acids

Carboxylic acids reacting as acids

Methanoic acid is present in stinging nettles and ant stings. Ethanoic acid, however, is the most well known as it is the main constituent of vinegar. Like other acids, carboxylic acids affect indicators and will react with metals such as magnesium. However, whereas the mineral acids, such as hydrochloric acid, are called strong acids, carboxylic acids are weak acids (there is more about weak acids in Chapter 8, p. 115). Even though they are weak acids, they will still react with bases to form salts. For example, the salt sodium methanoate (HCOONa) is formed when methanoic acid reacts with dilute sodium hydroxide.

methanoic + sodium → sodium + water
acid hydroxide methanoate

$$HCOOH(aq) + NaOH(aq) \rightarrow HCOONa(aq) + H_2O(l)$$

The second member of the homologous series, ethanoic acid, also undergoes typical reactions of acids with metals and carbonates. In the case of the metal magnesium, the metal salt magnesium ethanoate $((CH_3COO)_2Mg)$ and hydrogen are produced.

ethanoic + magnesium → magnesium + hydrogen
acid   ethanoate

$$2CH_3COOH(aq) + Mg(s) \rightarrow (CH_3COO)_2Mg(aq) + H_2(g)$$

Ethanoic acid reacts with carbonates such as sodium producing the salt sodium ethanoate (CH_3COONa), carbon dioxide and water.

ethanoic + sodium → sodium + carbon + water
acid carbonate ethanoate dioxide

$$2CH_3COOH(aq) + Na_2CO_3(s) \rightarrow 2CH_3COONa(aq) + CO_2 + H_2O(g)$$

Ethanoic acid also undergoes further typical reactions of acids with indicators turning, for example, methyl orange to red.

13.4 Esters

Ethanoic acid will react with ethanol, in the presence of a few drops of concentrated sulfuric acid acting as a catalyst, to produce ethyl ethanoate – an **ester**.

ethanoic + ethanol $\xrightleftharpoons[\text{H}_2\text{SO}_4]{\text{conc}}$ ethyl + water
acid   ethanoate

$$CH_3COOH(l) + C_2H_5OH(l) \xrightleftharpoons[\text{H}_2\text{SO}_4]{\text{conc}} CH_3COOC_2H_5(aq) + H_2O(l)$$

This reaction is called esterification.

Going further

Some interesting carboxylic acids

We come across carboxylic acids in our everyday life. Aspirin is a frequently used painkiller. It is also able to reduce inflammation and fever, and a low dose taken on a daily basis by people over the age of 50 may prevent heart attacks.

Vitamin C, also known as ascorbic acid, is an essential vitamin. Vitamin C is required by the body in very small amounts and it is obtained from foods. It is found in citrus fruits and brightly coloured vegetables, such as peppers and broccoli. Vitamin C prevents the disease scurvy.

Test yourself

6 Describe the reaction that takes place between ethanoic acid and:
   a calcium
   b potassium carbonate
   c potassium hydroxide.
   In each case construct a word equation as well as a balanced chemical equation.

7 Describe the reaction that takes place between the carboxylic acids, propanoic and butanoic, with:
   a magnesium
   b sodium carbonate.
   In each case, write a word equation, as well as constructing a balanced chemical equation.

Members of the 'ester' family have strong and pleasant smells. They have the general formula $C_nH_{2n+1}COOC_xH_{2x+1}$. Esters are named after the acid and alcohol from which they are derived:

» name – alcohol part first, acid part second, e.g. propyl ethanoate

» formula – acid part first, alcohol part second, e.g. $CH_3COOC_3H_7$

Table 13.6 shows the actual arrangement of the atoms in some of the members of this family.

▼ **Table 13.6** The displayed formulae of some esters

Ester	Made from		Structure	3D model
	Alcohol	Carboxylic acid		
Ethyl ethanoate $CH_3COOC_2H_5$	Ethanol C_2H_5OH	Ethanoic acid CH_3COOH		
Propyl methanoate $HCOOC_3H_7$	Propan 1-ol C_3H_7OH	Methanoic acid $HCOOH$		
Methyl butanoate $C_3H_7COOCH_3$	Methanol CH_3OH	Butanoic acid C_3H_7COOH		

Many esters occur naturally and are responsible for the flavours in fruits and the smells of flowers. They are used, therefore, in some food flavourings and in perfumes (Figure 13.10).

Fats and oils are naturally occurring esters which are used as energy storage compounds by plants and animals. They possess the same linkage as PET but have different units (see p. 208).

▲ **Figure 13.10** Perfumes contain esters

Test yourself

8 Which of the following organic chemicals are carboxylic acids or alcohols?
- Hexanoic acid, $C_5H_{11}COOH$
- Butanol, C_4H_9OH
- Octane, C_8H_{18}
- Nonane, C_9H_{20}
- Methanoic acid, $HCOOH$
- Pentene, C_5H_{10}
- Hexanol, $C_6H_{13}OH$

State why you have chosen your answers.

9 Write word and balanced chemical equations for the esterification of propanoic acid with ethanol.

Practical skills

Safety

● Eye protection must be worn.

A student wrote the following account of two experiments they observed their teacher carrying out about the properties of carboxylic acids.

1 Some <u>ethanoic acid</u> was poured into a test tube. A small piece of magnesium ribbon was added to the acid. <u>Bubbling</u> took place. The test tube was corked and the gas produced was tested using a lit wooden splint. It gave a squeaky 'pop'. The gas <u>must be oxygen</u>.

2 Some <u>ethanoic acid</u> was poured into a test tube. Some sodium carbonate was <u>added</u> to the acid. <u>Bubbling</u> took place.The test tube was corked and after about 10 seconds the cork was removed and the gas was poured into a test tube containing limewater. This test tube was corked <u>and shaken</u>. The limewater went slightly milky. The gas <u>must be hydrogen</u>.

Some statements in the student's account have been underlined by the teacher.

Answer these questions to identify how the account of the experiments could be improved.

1 Which piece of apparatus would you use to measure out 10 cm^3 of the acid?
2 In both cases, what strength of ethanoic acid, dilute or concentrated, and volume of acid, 5 cm^3 or 50 cm^3, would be used?
3 In both experiments, what is the more scientifically correct word to use rather than 'bubbling'?
4 What was the gas that gave a squeaky pop?
5 In experiment 2, how is the sodium carbonate added?
6 In experiment 2, how should the shaking take place? Explain your answer.
7 In experiment 2, what was the gas that gave a milky colour with limewater?
8 Are there any other suggestions that you could make to improve the account?

13.5 Condensation polymers

In Chapter 12 (p. 191) you studied the different addition polymers produced from alkenes. Not all polymers are formed by addition reactions though. Some are produced as a result of a different type of reaction. In 1935, Wallace Carothers discovered a different sort of plastic when he developed the thermoplastic, nylon. Nylon is made by reacting two different chemicals together, unlike poly(ethene) which is made only from monomer units of ethene. Poly(ethene), formed by addition polymerisation, can be represented by:

–A–A–A–A–A–A–A–A–A–A–

where A = monomer.

The starting molecules for nylon are more complicated than those for poly(ethene) and are called 1,6-diaminohexane and hexanedioic acid.

Hexanedioic acid $\qquad$ + $\qquad$ 1,6-diaminohexane
$HOOC(CH_2)_4COOH$ $\qquad\qquad$ $H_2N(CH_2)_6NH_2$

$H_2N(CH_2)_6\mathbf{CONH}(CH_2)_4COOH$ + H_2O

Amide link

The polymer chain is made up from the two starting molecules arranged alternately (Figure 13.11) due to these molecules reacting and therefore linking up. Each time a reaction takes place, the other product of the reaction a molecule of water, is lost.

▲ **Figure 13.11** A nylon polymer chain is made up from the two molecules arranged alternately just like the two different coloured poppet beads in the photo

This sort of reaction is called **condensation polymerisation**. This differs from addition polymerisation, where there is only one product. Because an amide link is formed during the polymerisation, nylon is known as a **polyamide**. This is the same amide link as found in proteins (p. 209). It is often called the peptide link. This type of polymerisation, in which two kinds of monomer unit react, results in a chain of the type:

−A−B−A−B−A−B−A−B−A−B−

Generally, polyamides have the structure:

$$-\overset{O}{\underset{||}{C}}-\blacksquare-\overset{O}{\underset{||}{C}}-\underset{H}{N}-\square-\underset{H}{N}-\overset{O}{\underset{||}{C}}-\blacksquare-\overset{O}{\underset{||}{C}}-\underset{H}{N}-\square-\underset{H}{N}-\overset{O}{\underset{||}{C}}-\blacksquare$$

Looking at this structure it is easy to see what the repeat unit of this polyamide is.

When nylon is made in industry, it forms as a solid which is melted and forced through small holes (Figure 13.12). The long filaments cool and solid nylon fibres are produced, which are stretched to align the polymer molecules and then dried. The resulting yarn can be woven into fabric to make shirts, ties, sheets and parachutes or turned into ropes or racket strings for tennis and badminton rackets.

▲ **Figure 13.12** Nylon fibre is formed by forcing molten plastic through hundreds of tiny holes

We can obtain different polymers with different properties if we carry out condensation polymerisation reactions between other monomer molecules. For example, if we react ethane-1,2-diol with benzene-1,4-dicarboxylic acid (terephthalic acid), then we produce a polymer called poly(ethyleneterephthalate) or PET for short.

Ethane-1,2-diol $\qquad$ + $\qquad$ Benzene-1,4-dicarboxylic acid
$HO(CH_2)_2OH$ $\qquad\qquad$ $HOOC(C_6H_4)COOH$

$$\downarrow$$

$$HO(CH_2)_2\mathbf{OCO}(C_6H_4)COOH + H_2O$$

Ester link

This ester link is the same linkage as in fats. Generally, polyesters have the structure:

$$-\overset{O}{\underset{||}{C}}-\blacksquare-\overset{O}{\underset{||}{C}}-O-\square-O-\overset{O}{\underset{||}{C}}-\blacksquare-\overset{O}{\underset{||}{C}}-O-\square-O-$$

Like nylon, PET can be turned into yarn, which can then be woven. PET clothing is generally softer than that made from nylon but both are hard wearing. Because an ester link is formed during the polymerisation, PET is known as a **polyester**. When this polymer is being used to make bottles, for example, it is usually called PET. An interesting feature of PET is that it can be converted back into its monomers and so can be recycled.

> **Test yourself**
>
> 10 Draw the displayed structures of the two monomer units for the following polymer chain:
>
>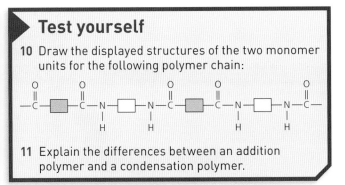
>
> 11 Explain the differences between an addition polymer and a condensation polymer.

13.6 Natural polyamides

Proteins are natural polyamides. They are condensation polymers formed from amino acid monomers. Amino acid monomers have the general structure shown below. R represents different types of side chain.

There are 20 different amino acids and they each possess two functional groups. One is the carboxylic acid group, –COOH. The other is the amine group, –NH$_2$. Two amino acids are glycine and alanine.

General structure   Glycine   Alanine

Amino acids are the building blocks of proteins. Similar to nylon (see p. 207), proteins are

polyamides as they contain the –CONH– group, which is called the amide or, in the case of proteins, the peptide link. Proteins are formed by condensation polymerisation.

Glycine   Alanine

a **dipeptide**
(composed of two amino acids joined together)

Protein chains formed by the reaction of many amino acid molecules have the general structure shown below.

Further reaction with many more amino acids takes place at each end of each molecule to produce the final protein (Figures 13.13 and 13.14). For a molecule to be a protein, there must be at least 100 amino acids involved. Proteins make up 15% of our body weight and, along with fats and carbohydrates (such as starch and glucose), they are among the main constituents of food.

▲ **Figure 13.13** General structure of a protein

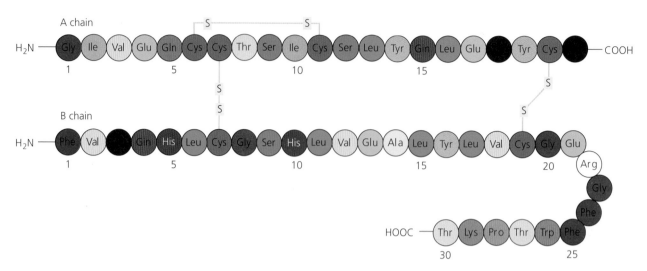

▲ **Figure 13.14** The structure of a protein – human insulin (the different coloured circles represent different amino acids in this protein)

Test yourself

12 How many amino acids have to be involved before the biopolymer is called a protein?

13 a Give the molecular formula of the amino acid valine, shown here.

b During a condensation reaction involving this amino acid which two functional groups react?

Revision checklist

After studying Chapter 13 you should be able to:

✔ Write the general formulae for the homologous series, alcohols and carboxylic acids.

✔ Name and draw the structural and displayed formulae of unbranched alcohols, including propan-1-ol, propan-2-ol, butan-1-ol and butan-2ol, and carboxylic acids, as well as the products of their reactions containing up to four carbon atoms per molecule.

✔ State the type of compound present given the chemical name ending in -ol or -oic acid or from a molecular, structural or displayed formula.

✔ Describe the manufacture of ethanol by fermentation of glucose and the catalytic addition of steam to ethene.

✔ Describe the advantages and disadvantages of the two methods for manufacturing ethanol.

✔ Describe the combustion of alcohols.

✔ State the uses of ethanol as a solvent and a fuel.

✔ Describe the reactions of carboxylic acids with metals, bases and carbonates including names and formulae of the salts produced.

✔ Describe the formation of ethanoic acid by the oxidation of ethanol with acidified potassium manganate(VII) as well as by bacterial oxidation during vinegar production.

✔ Describe the reaction of a carboxylic acid with an alcohol to give an ester.

✔ Name and draw the displayed formulae of the unbranched esters which can be made from unbranched alcohols and carboxylic acids, each containing up to four carbon atoms.

✔ Deduce the structure or repeat unit of a condensation polymer from given monomers.

✔ Describe the differences between condensation and addition polymerisation.

✔ Describe and draw the structure of nylon (a polyamide) and PET (a polyester).

✔ Describe proteins as natural polyamides.

✔ Draw the general structure of a natural polyamide.

✔ Describe and draw the structure of proteins.

Exam-style questions

1 a The *functional group* of the *homologous series* alcohols is –OH. Explain what is meant by the terms in italics. [4]

b Give the general formula of the homologous series of alcohols. [1]

c Butan-1-ol is used as a solvent for paints. It can be made by the hydration of but-1-ene.

 i Give an equation for the reaction that takes place to produce butan-1-ol. [3]

 ii Identify the reaction conditions for this reaction to take place. [3]

d Butan-1-ol will undergo bacterial oxidation if left in the air. Give the name of the organic substance formed by this oxidation process and give its formula. [2]

2 A piece of cheese contains protein. Proteins are natural polymers made up of amino acids. There are 20 naturally occurring amino acids. The displayed formulae of two amino acids are shown below.

Glycine   Alanine

$$H_2N-\underset{\underset{H}{|}}{\overset{\overset{H}{|}}{C}}-COOH \qquad H_2N-\underset{\underset{CH_3}{|}}{\overset{\overset{H}{|}}{C}}-COOH$$

a Identify the type of polymerisation involved in protein formation. [1]

b Give the molecular formulae of the two amino acids shown. [2]

c Sketch a displayed formula to represent the product formed by the reaction between the amino acids shown above. [2]

3 a Copy the following table and complete it by giving the displayed formulae for ethanol and ethanoic acid.

Ethane	Ethanol	Ethanoic acid
$H-\underset{\underset{H}{|}}{\overset{\overset{H}{|}}{C}}-\underset{\underset{H}{|}}{\overset{\overset{H}{|}}{C}}-H$		

[4]

b Describe a simple chemical test that could be used to distinguish ethanol from ethanoic acid. [3]

c i Identify the class of compound produced when ethanol reacts with ethanoic acid. [1]

 ii What catalyst is required for this reaction? [1]

 iii Give a word and balanced chemical equation for this reaction. [4]

4 a Ethanoic acid is a weak acid. Describe what you understand by the term weak acid. [1]

b Ethanoic acid will react with sodium carbonate. Write a word and balanced chemical equation for the reaction. [4]

c Ethanoic acid will react with metals such as magnesium. Write a word and balanced chemical equation for the reaction. [4]

d Ethanoic acid will react with bases such as sodium hydroxide.

 i What do you understand by the term 'base'? [2]

 ii Give another example of a base. [1]

 iii Write a word and balanced chemical equation for the reaction. [4]

5 a Identify which carboxylic acids and alcohols you use to make the following esters:

 i butyl methanoate [2]

 ii ethyl methanoate [2]

 iii propyl propanoate. [2]

b Identify which esters would be made using the following carboxylic acids and alcohols:

 i ethanoic acid and butan-1-ol [1]

 ii methanoic acid and methanol [1]

 iii propanoic acid and butan-1-ol. [1]

c Sketch the displayed formulae of the following esters:

 i ethyl ethanoate [2]

 ii propyl ethanoate. [2]

6 a Identify the polymerisation process that is used to make both nylon and PET. [1]

b Give the starting materials for making:

 i nylon [2]

 ii PET. [2]

c Give the name and formula of the small molecule produced during the polymerisation reactions to produce both nylon and PET. [2]

d Give the name of the chemical link that holds together:

 i nylon [1]

 ii PET. [1]

Experimental techniques and chemical analysis

This chapter shows how to safely and accurately use different experimental techniques, apparatus and materials. This will aid your ability to plan experiments and investigations.

You will also learn that there are specific criteria for purity of substances. Finally, you will learn about the branch of chemistry that deals with the identification of elements, ions or grouping of elements as ions present in a sample, which is called qualitative chemical analysis, or qualitative analysis for short. The various tests you will study would allow you to carry out a wide range of qualitative analysis.

14.1 Apparatus used for measurement in chemistry

Scientists find out about the nature of materials by carrying out experiments in a laboratory. In your course, you have used scientific apparatus to make measurements in experiments. A knowledge and understanding of the correct use of this scientific apparatus is required for successful experimentation and planning investigations. In an experiment, you will first have to decide on the measurements to be made and then collect together the apparatus and materials required. The quantities you will need to measure most often in a chemistry laboratory will be temperature, mass, volume of liquids as well as gases, and time.

There will be advantages and disadvantages of each method or piece of apparatus, so when you choose which to use you will have to use your knowledge of the apparatus to decide if it will be suitable. You will have to decide:

>> What apparatus should you use to measure each of these? For example, it sounds silly but you would not use a thermometer to measure the timing of a particular reaction rate!
>> Which piece of apparatus is most suitable for the task? For example, you would not use a large gas syringe to measure out 10 cm³ of liquid: you would use a small measuring cylinder because

it gives you a more accurate measurement of volume and it is easier to use.
>> How do you use the piece of apparatus correctly? Before you use a piece of apparatus, be sure you know how to use it properly. You need to be safe in your working habits and also to ensure you use good techniques, for example measuring volume of solutions carefully, so that you make and record measurements accurately. If necessary, ask your teacher before you use a particular piece of apparatus if it is unfamiliar to you.

Measurement of time

▲ **Figure 14.1** This stopwatch can be used to measure the time passed in a chemical reaction. The reading is 6.3 s

Experiments involving rates of reaction will require the use of a stopwatch (Figure 14.1) that measures to a hundredth of a second. The units of time are hours (h), minutes (min) and seconds (s).

Measurement of temperature

▲ **Figure 14.2** A thermometer can be used to measure temperature. The reading is 3°C

The most commonly used thermometers in a laboratory are alcohol-in-glass. Mercury in-glass thermometers can be used but should be handled with great care. The mercury inside them is poisonous and should not be handled if a thermometer breaks. The units of temperature are those of the Celsius scale. This scale is based on the temperature at which water freezes and boils, that is:

» the freezing point of water is 0°C
» the boiling point of water is 100°C.

The usual thermometer used is that shown in Figure 14.2, which measures accurately between –10° and 110°C at 1°C intervals. It may also be possible to read the thermometer to the nearest 0.5°C if the reading is between two of the scale marks. When reading the thermometer always ensure that your eye is at the same level as the liquid meniscus in the thermometer to ensure there are no parallax effects (Figure 14.3). The meniscus is the way that the liquid curves at the edges of the capillary in which the liquid is held in the thermometer.

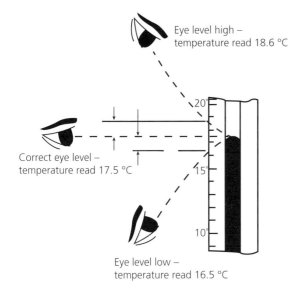

Eye level high – temperature read 18.6 °C

Correct eye level – temperature read 17.5 °C

Eye level low – temperature read 16.5 °C

▲ **Figure 14.3** Eye should be level with the liquid meniscus

Measurement of mass

▲ **Figure 14.4** An electronic balance can be used to measure the mass of reagents. The reading is 155.12 g

There are many different electronic balances (Figure 14.4) which can be used. The precision of an electronic balance is the size of the smallest mass that can be measured on the scale setting you are using, usually 0.01 g. The units for measuring mass are grams (g) and kilograms (kg): 1 kg = 1000 g.

When using an electronic balance you should wait until the reading is steady before taking it.

Measurement of volume of liquids

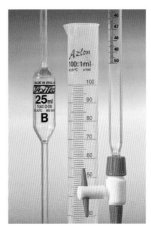

▲ **Figure 14.5** The apparatus shown in the photograph (left to right): volumetric pipette, measuring cylinder and burette, are generally used in different experiments to measure volume accurately

Different experiments involving liquids will require one or other or all the various measuring apparatus available for measuring volume. Figure 14.5 shows three of the most commonly used measuring apparatus for measuring volume of a liquid.

Acid-base titration

A burette and a **pipette** are needed in a titration. This technique is often used to test the strength or purity of an acid or an alkali. A titration involves the slow, careful addition of one solution, such as an acid, of a known concentration from a burette to a known volume of another solution, such as an alkali, of unknown concentration held in a flask. The solution in the burette is known as a **titrant**. This addition of acid continues until the reaction just reaches neutralisation. The end-point of a titration is indicated by a colour change in an indicator. For a further discussion of indicators, and identifying the end-point, see Chapter 8 (p. 119). The known volume in the flask would require the use of a pipette. For example, in an acid-alkali titration, the acid would be the titrant. For a further discussion of titrations, see Chapter 8, p. 124.

A measuring cylinder (sometimes called a mixing or graduated cylinder) is a common piece of laboratory equipment used to measure the volume of a liquid. It has a narrow cylindrical shape. Each marked line on the graduated cylinder shows the amount of liquid that has been measured.

The volume of a liquid is a measure of the amount of space that it takes up. The units of volume are litres (l), cubic decimetres (dm^3) and cubic centimetres (cm^3).

$$1 \text{ litre} = 1 \text{ dm}^3 = 1000 \text{ cm}^3$$

However, some of the manufacturers of apparatus used for measuring volume use millilitres (ml). This is not a problem, however, since $1 \text{ cm}^3 = 1 \text{ ml}$.

When reading the volume using one of these pieces of apparatus, it is important to ensure that the apparatus is vertical and that your eye is level with the meniscus of the liquid being measured, as shown in Figure 14.6. The precision of the measurement will vary depending on the apparatus used. For example, if the smallest scale division on a burette is 0.1 cm^3, the precision of any volume

measured with the burette will be 0.1 cm^3. It may also be possible to read the burette to the nearest 0.05 cm^3 if the bottom of the meniscus is between divisions.

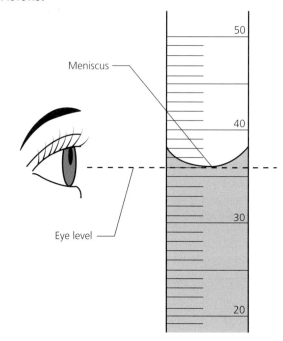

▲ **Figure 14.6** The volume level in this measuring cylinder should be read on the dotted line. The reading here is 36 cm^3

Measurement of volume of gases

The volume of a gas can be measured with a gas syringe. This is used to measure the amount of gas collected in experiments. They have a maximum volume of 100 cm^3 (Figure 14.7).

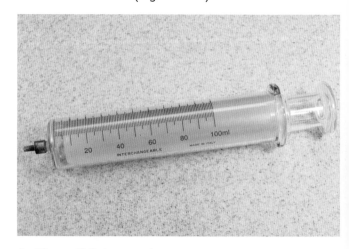

▲ **Figure 14.7** A gas syringe

For example, in certain reactions (see p. 102) the reaction rate can be followed by collecting the volume of gas generated with time. The gas syringe is connected by a tube to the flask that is giving off the gas as shown in Figure 14.8.

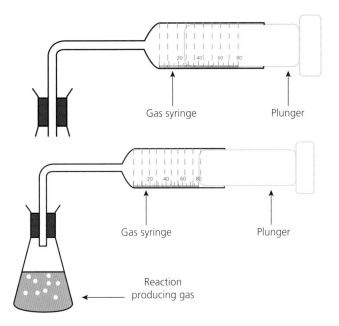

Gas syringe   Plunger

Gas syringe   Plunger

Reaction producing gas

▲ **Figure 14.8** As gas is collected during the reaction, the plunger is forced out and the volume can be read from the scale on the side

The gas syringe has a scale along its length that allows the volume of the gas collected to be measured. As the volume of gas collected increases, the plunger moves further out of the syringe. The volume of collected gas can be measured in a given time, as it increases, and a graph of volume against time can be plotted. This can then be used to show the rate of the reaction (see Chapter 7, p. 102).

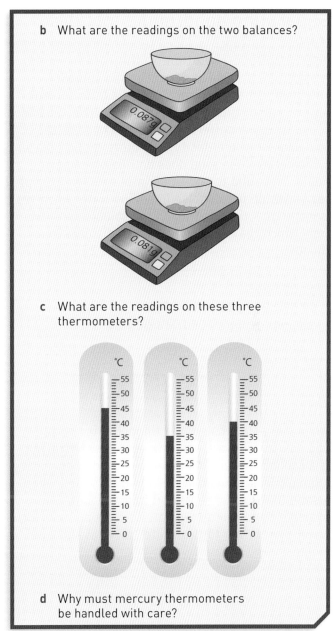

b What are the readings on the two balances?

c What are the readings on these three thermometers?

d Why must mercury thermometers be handled with care?

14.2 Separating mixtures

Many mixtures contain useful substances mixed with unwanted material. To obtain these useful substances, chemists often have to separate them from the impurities. Chemists have developed many different methods of separation, particularly for separating compounds from complex mixtures. Which separation method they use depends on what is in the mixture and the properties of the substances present. It also depends on whether the substances to be separated are solids, liquids or gases.

> **Test yourself**
>
> 1 **a i** Copy and complete the following about the units of volume measurement.
> $1 \, dm^3 =$ _____ cm^3
>   **ii** What are the readings on the following measuring cylinders?

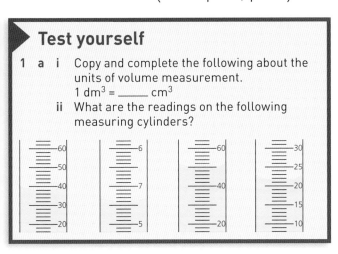

Separating solid/liquid mixtures

If a solid substance is added to a liquid it may dissolve to form a **solution**. In this case the solid is said to be soluble and is called the **solute**. The liquid it has dissolved in is called the **solvent**. An example of this type of process is when sugar is added to tea or coffee. What other examples can you think of where this type of process takes place?

Sometimes the solid does not dissolve in the liquid. This solid is said to be insoluble. For example, tea leaves do not dissolve in boiling water when tea is made from them, although the soluble materials from which tea is made are seen to dissolve from them.

> **Key definitions**
>
> A **solvent** is a substance that dissolves a solute.
>
> A **solute** is a substance that is dissolved in a solvent.
>
> A **solution** is a liquid mixture composed of two or more substances.

Filtration

When you pour a cup of tea through a tea strainer you are carrying out a filtering process. **Filtration** is a common separation technique used in chemistry laboratories throughout the world. It is used when a solid needs to be separated from a liquid. For example, sand can be separated from a mixture with water by filtering through filter paper as shown in Figure 14.9.

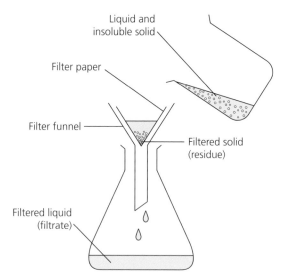

Labels: Liquid and insoluble solid; Filter paper; Filter funnel; Filtered solid (residue); Filtered liquid (filtrate)

▲ **Figure 14.9** It is important when filtering not to overfill the filter paper

The filter paper contains holes that, although too small to be seen, are large enough to allow the molecules of water through but not the sand particles. It acts like a sieve. The sand gets trapped in the filter paper and the water passes through it. The sand is called the **residue** and the water is called the **filtrate**.

> **Key definitions**
>
> **Residue** is a substance that remains after evaporation, distillation, filtration or any similar process.
>
> **Filtrate** is a liquid or solution that has passed through a filter.

Going further

Decanting

Vegetables do not dissolve in water. When you have boiled some vegetables it is easy to separate them from the water by pouring it off. This process is called decanting. This technique is used quite often to separate an insoluble solid, which has settled at the bottom of a flask, from a liquid.

Centrifuging

Another way to separate a solid from a liquid is to use a centrifuge. This technique is sometimes used instead of filtration. It is usually used when the solid particles are so small that they spread out (disperse) throughout the liquid and remain in suspension. They do not settle to the bottom of the container, as heavier particles would do, under the force of gravity. The technique of centrifuging or centrifugation involves the suspension being spun round very fast in a centrifuge so that the solid gets flung to the bottom of the tube (Figure 14.10).

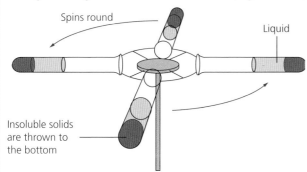

a The sample is spun round very fast and the solid is flung to the bottom of the tube

b An open centrifuge

▲ **Figure 14.10**

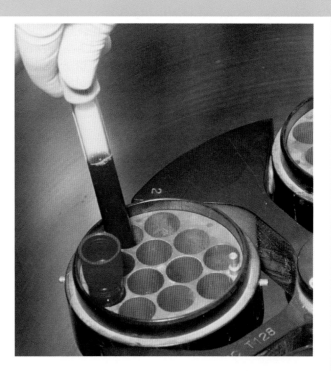

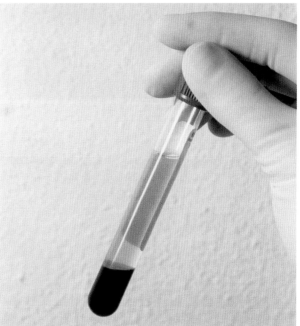

▲ **Figure 14.11** Whole blood (top) is separated by centrifuging into blood cells and plasma (bottom)

The pure liquid can be decanted after the solid has been forced to the bottom of the tube. This method of separation is used extensively to separate blood cells from blood plasma (Figure 14.11). In this case, the solid particles (the blood cells) are flung to the bottom of the tube, allowing the liquid plasma to be decanted.

Evaporation

If a solid has dissolved in a liquid it cannot be separated by filtering or centrifuging. Instead, the solution can be heated so that the liquid **evaporates** completely and leaves the solid behind. For example, the simplest way to obtain salt from its solution is by slow evaporation, as shown in Figure 14.12.

▲ **Figure 14.12** Apparatus used to slowly evaporate a solvent

It may be necessary to choose a suitable solvent. For example, you may have a mixture of substances of which one is soluble in water. This would be the suitable solvent to separate the soluble substance. Filtration could then be used followed by evaporation of the water. See also Solvent extraction on p. 221.

Crystallisation

In many parts of the world, salt is obtained from sea water on a vast scale. This is done by using the heat of the Sun to evaporate the water to leave a saturated solution of salt known as brine. A **saturated solution** is defined as one that contains the maximum concentration of a solute dissolved in the solvent. It is important to be aware that solubility of a substance varies with temperature, so the maximum concentration will vary at different temperatures. More of the solute will dissolve at a higher temperature. When the solution is saturated the salt begins to **crystallise** and it is removed using large scoops (Figure 14.13).

> **Key definition**
>
> A **saturated solution** is a solution containing the maximum concentration of a solute dissolved in the solvent at a specified temperature.

▲ **Figure 14.13** Salt is obtained in north-eastern Brazil by evaporation of sea water

Simple distillation

If we want to obtain the solvent from a solution, then the process of **distillation** can be carried out. The apparatus used in this process is shown in Figure 14.14.

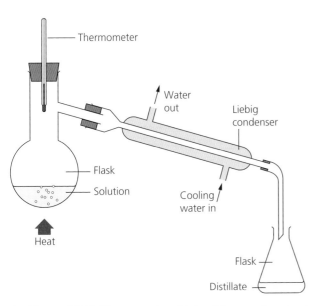

▲ **Figure 14.14** Water can be obtained from salt water by distillation

Water can be obtained from salt water using this method. The solution is heated in the flask until it boils. The steam rises into the Liebig condenser, where it condenses back into water. The salt is left behind in the flask.

➡ **Going further**

Desalination

In hot and arid countries, such as Saudi Arabia, distillation is used on a much larger scale to obtain pure water for drinking (Figure 14.15). This process is carried out in a desalination plant.

▲ **Figure 14.15** This desalination plant produces large quantities of drinking water in Saudi Arabia

Separating liquid/liquid mixtures

In recent years there have been several oil tanker accidents that have led to environmental disasters, just like the one shown in Figure 14.16. These have resulted in millions of litres of oil being washed into the sea. Oil and water do not mix easily. They are said to be **immiscible**. When cleaning up disasters of this type, a range of chemicals can be added to the oil to make it more soluble. This results in the oil and water mixing with each other. They are now said to be **miscible**. The following techniques can be used to separate mixtures of liquids.

▲ **Figure 14.16** Millions of litres of oil can be spilled from oil tankers and cleaning up is a slow and costly process

Immiscible liquids

If two liquids are immiscible they can be separated using a separating funnel. The mixture is poured into the funnel and the layers allowed to separate. The lower layer can then be run off by opening the tap as shown in Figure 14.17.

▲ **Figure 14.17** The blue liquid is more dense than the red liquid and so sinks to the bottom of the separating funnel. When the tap is opened the blue liquid can be run off.

Miscible liquids

If miscible liquids are to be separated, then this can be done by fractional distillation. The apparatus used for this process is shown in Figure 14.18 and could be used to separate a mixture of ethanol and water.

Fractional distillation relies upon the liquids having different boiling points. When an ethanol and water mixture is heated, the vapours of ethanol and water boil off at different temperatures and can be condensed and collected separately.

Ethanol boils at 78°C whereas water boils at 100°C. When the mixture is heated the vapour produced is mainly ethanol with some steam. Because water has the higher boiling point of the two, it condenses out from the mixture with ethanol. This is what takes place in the fractionating column. The water condenses and drips back into the flask while the ethanol vapour moves up the column and into the condenser, where it condenses into liquid ethanol and is collected in the receiving flask as the **distillate**. When all the ethanol has distilled over, the temperature reading on the thermometer rises steadily to 100°C, showing that the steam is now entering the condenser. At this point the receiver can be changed and the condensing water can now be collected.

Fractional distillation is used to separate miscible liquids such as those in petroleum (crude oil) (see p. 85, Figure 6.4 and b), and the technique can also separate individual gases, such as nitrogen, from the mixture we call air (Figure 14.19).

Thermometer

Condenser

Cooling water in

Cooling water out

Fractioning column with short lengths of glass rod inside (increase surface area)

Flask

Liquid mixture

Heat

Flask

Distillate

Support

▲ **Figure 14.18** Typical fractional distillation apparatus

▲ **Figure 14.19** Gases from the air are extracted in this fractional distillation plant

→ Going further

Separating solid/solid mixtures

You saw in Chapter 2 (Figure 2.9, p. 18) that it was possible to separate iron from sulfur using a magnet. In that case we were using one of the physical properties of iron, that is, the fact that it is magnetic. In a similar way, it is possible to separate scrap iron from other metals by using a large electromagnet like the one shown in Figure 14.20.

▲ **Figure 14.20** Magnetic separation of iron-containing materials

It is essential that when separating solid/solid mixtures you pay particular attention to the individual physical properties of the components. If, for example, you wish to separate two solids, one of which sublimes, then this property should dictate the method you employ.

In the case of an iodine/common salt mixture, the iodine sublimes but salt does not. Iodine can be separated by heating the mixture in a fume cupboard as shown in Figure 14.21. The iodine sublimes and re-forms on the cool inverted funnel.

▲ **Figure 14.21** Apparatus used to separate an iodine/salt mixture. The iodine sublimes on heating

Solvent extraction

Sugar can be obtained from crushed sugar cane by adding water. The water dissolves the sugar from the sugar cane (Figure 14.22). This is an example of solvent extraction. In a similar way, some of the green substances can be removed from ground-up grass using ethanol. The substances are extracted from a mixture by using a solvent which dissolves only those substances required.

▲ **Figure 14.22** Cutting sugar cane, from which sugar can be extracted, by using a suitable solvent

> ## Test yourself
>
> 2 Which experimental technique(s)/processes would you use to obtain:
>   a the silver bromide precipitate produced from the reaction of silver nitrate solution with sodium bromide solution?
>   b pure water from an aqueous solution of iron(II) sulfate?
>   c common salt from a mixture of the solids salt and sand?
>   d nitrogen gas (boiling point −195.8°C) from liquid air?
>   e calcium chloride solid from calcium chloride solution?

Chromatography

Sometimes we need to separate two or more solids that are soluble, for example mixtures of soluble coloured substances such as inks and dyes. A technique called **chromatography** is widely used to separate these materials so that they can be identified.

There are several types of chromatography; however, they all follow the same basic principles. The simplest kind is paper chromatography. To separate the different coloured dyes in a sample of black ink, a pencil line is drawn about 1 cm from the bottom of a piece of chromatography paper. This is called the baseline. Do not use a pen to draw the baseline as the dyes in its ink will interfere with the results. A spot of ink is put onto the pencil line on the chromatography paper. This paper is then put into suitable container, such as a beaker, containing a suitable solvent, ensuring that the solvent level does not reach the pencil line as shown in Figure 14.23. A lid is placed over the container to reduce evaporation of the solvent.

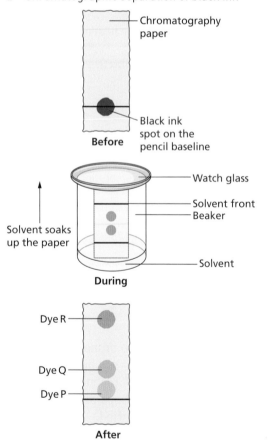

a Chromatographic separation of black ink

b The black ink separates into three dyes: P, Q and R

▲ **Figure 14.23**

As the solvent moves up the paper, the dyes are carried with it and begin to separate. They separate because the substances have different solubilities

in the solvent and are absorbed to different degrees by the chromatography paper. As a result, they are separated gradually as the solvent moves up the paper. The **chromatogram** in Figure 14.23b shows how the ink contains three dyes, P, Q and R.

Chromatography is used extensively in medical research and forensic science laboratories to separate a variety of mixtures.

? Worked example

Several different dyes were spotted on chromatography paper and set up for chromatography. The chromatogram produced is shown below.

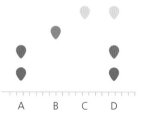

Which of the dyes are pure substances and which are mixtures?

The single spots show they are a single dye. So dyes B and C are single dyes.

The dyes that have produced multiple spots are mixtures. So dyes A and D are mixtures of dyes.

⬇ Going further

Paper chromatography of the products of starch hydrolysis

This technique can be used to identify the products of a reaction. For example, starch can be broken up using enzymes in water solution or water in dilute acid. This type of reaction is called hydrolysis. However, the different methods produce different products, as shown in Figure 14.24.

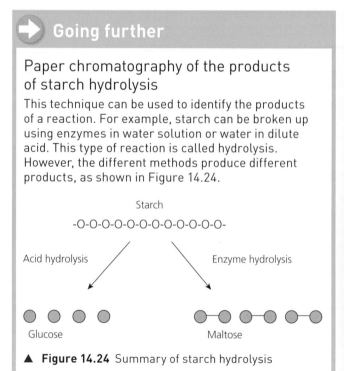

▲ **Figure 14.24** Summary of starch hydrolysis

The products of these reactions can be identified using chromatography. The chromatography paper is 'spotted' with the samples as well as pure glucose and maltose. When this is developed after some hours in the solvent, the spots may be identified (Figure 14.25).

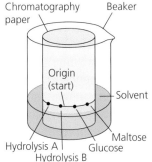

Chromatography paper   Beaker

Origin (start)

Solvent

Maltose
Hydrolysis A   Glucose
Hydrolysis B

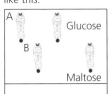

After several hours in the solvent, the chromatogram is sprayed with a substance that makes the sugars stand out. It then looks like this.

A   Glucose
B
Maltose

▲ **Figure 14.25** The products of hydrolysis may be identified by chromatography

? Worked example

Which substance has been produced by which hydrolysis in Figure 14.25?

Acid hydrolysis – sample A on the chromatogram. That spot has risen to the same height as the substance glucose.

Enzyme hydrolysis – sample B on the chromatogram. That spot has risen to the same height as the substance maltose.

Numerical measurements (retardation factors) known as **R_f values** can be obtained from chromatograms. An R_f value is defined as the ratio of the distance travelled by the solute (for example P, Q or R) to the distance travelled by the solvent from the pencil line.

$$R_f = \frac{\text{distance travelled by solute}}{\text{distance travelled by solvent}}$$

The substances to be separated do not have to be coloured. Colourless substances can be made visible by spraying the chromatogram with a **locating agent**. The locating agent will react with the colourless substances to form a coloured product. In other situations, the position of the substances on the chromatogram may be located using ultraviolet light.

? Worked example

The following values were obtained for the chromatogram produced for the amino acid glycine.

Distance travelled by the glycine = 2.5 cm

Distance travelled by the solvent = 10.0 cm

$$R_f = \frac{\text{distance travelled by solute}}{\text{distance travelled by solvent}}$$

$$R_f = \frac{2.5}{10.0} = 0.25$$

R_f values are catalogued in data books for a vast array of substances. So R_f values of unknown substances can be compared with values inside a data book to identify the unknown substance.

Criteria for purity

In chemistry, a pure substance is one which contains a single element or compound not mixed with any other substance. The measure of whether a substance is pure is known as purity. This is different to the everyday meaning. For example, in the food industry, if bottles or cartons which are labelled as 'pure' apple juice are sold in shops then the label means that the contents are just apple juice with no other substances being added. However, the apple juice is not pure in the chemical sense because it contains different substances mixed together. The substances added here may include a preservative as well as other substances which are harmless to the person buying it.

However, in the pharmaceuticals industry, purity is essential because there could be significant dangers to health if any other substances were found in medicines. Drugs (pharmaceuticals) are manufactured to a very high degree of purity (Figure 14.26). To ensure that the highest possible purity is obtained, the drugs are dissolved in a suitable solvent and subjected to fractional crystallisation.

Purity can have an important effect on the chemical properties of a substance. Pure substances will form predictable products in chemical reactions. This reliability comes from the fact that there are no impurities present that could start other reactions. Reactions due to impurities may interfere with the reaction that is being studied by a chemist, so chemists often use substances which have a high degree of purity when conducting chemical research.

Throughout the pharmaceutical, food and chemical industries, it is essential that the substances used are pure. The purity of a substance can be assessed by:

» Its melting point – if it is a pure solid it will have a sharp melting point. If an impurity is present then melting takes place over a range of temperatures to show it is a mixture of two or more substances.
» Its boiling point – if it is a pure liquid the temperature will remain steady at its boiling point. If the substance is impure then the mixture will boil over a temperature range.
» Chromatography – if it is a pure substance it will produce only one well-defined spot on a chromatogram. If impurities are present then several spots will be seen on the chromatogram in Figure 14.23, p. 222.

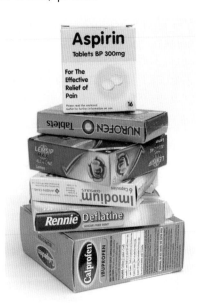

▲ **Figure 14.26** These pharmaceuticals must have been through a lot of testing for purity before they can be sold in a pharmacy

> **Test yourself**
>
> 3 The following values were obtained from the chromatogram produced for the amino acid proline.
>
>   Distance travelled by proline = 6.0 cm
>
>   Distance travelled by the solvent = 10.0 cm
>
>   Calculate the R_f value for proline.

14.3 Qualitative analysis

The branch of chemistry that deals with the identification of elements or grouping of elements present in a sample is called qualitative chemical analysis, or qualitative analysis for short. It does not deal with quantities.

The techniques employed in qualitative analysis vary in their complexity, depending on the nature of the sample under investigation. In some cases, it is only necessary to confirm the presence of certain elements or groups for which specific chemical tests applicable directly to the sample may be available. More often, the sample is a complex mixture, and a systematic analysis must be made in order that all the component parts may be identified. Often, the first simple stages of qualitative analysis require no apparatus at all. Things like colour and smell can be observed without any need for apparatus.

The following summary collects together information from throughout the book which would allow you to carry out qualitative analysis. The material which is of extreme importance as far as your revision is concerned is that given in the following tables:

» Table 14.2 Characteristic flame colours of some metal ions
» Table 14.3 Effect of adding sodium hydroxide solution to solutions containing various metal ions
» Table 14.4 Effect of adding aqueous ammonia to solutions containing various metal ions
» Table 14.5 A variety of tests for aqueous anions
» Table 14.6 Tests for gases

Appearance or smell

A preliminary examination of the substance will give you a start. The appearance or smell of a substance can often indicate what it might contain (Table 14.1).

▼ **Table 14.1** Deductions that can be made from a substance's appearance or smell

Observation on substance	Indication
Black powder	Carbon, or contains O^{2-} ions (as in CuO), or S^{2-} ions (as in CuS)
Pale green crystals	Contains Fe^{2+} ions (as in iron(II) salts)
Dark green crystals	Contains Ni^{2+} ions (as in nickel(II) salts)
Blue or blue-green crystals	Contains Cu^{2+} ions (as in copper(II) salts)
Yellow-brown crystals	Contains Fe^{3+} ions (as in iron(III) salts)
Smell of ammonia	Contains NH_4^+ ions (as in ammonium salts)

Flame colours

If a wooden splint, which has been soaked in an aqueous metal ion solution, is held in a colourless Bunsen flame, the flame colour can become coloured (Figure 14.27). Certain metal ions may be detected in their compounds by observing their flame colours (Table 14.2).

▲ **Figure 14.27** The green colour is characteristic of copper

▼ **Table 14.2** Characteristic flame colours of some metal ions

	Metal	Flame colour
Group I (1+ ion)	Lithium	Red
	Sodium	Yellow
	Potassium	Lilac
Group II (2+ ion)	Calcium	Orange-red
	Barium	Light green
Others	Copper (as Cu^{2+})	Blue-green

A flame colour is obtained as a result of the electrons in the particular ions being excited when they absorb energy from the flame which is then emitted as visible light. The different electronic configurations of the different ions, therefore, give rise to the different colours.

Tests for aqueous cations

Effect of adding dilute sodium hydroxide solution

Aqueous sodium hydroxide can be used to identify salts of Al^{3+}, Ca^{2+}, Cr^{3+}, Cu^{2+}, Fe^{2+}, Fe^{3+} and Zn^{2+} when present in aqueous solutions. All metal cations form insoluble hydroxides when sodium hydroxide solution is added to them. The colour of the precipitate and its behaviour in excess sodium hydroxide solution will help identify the metal present (Table 14.3).

▼ **Table 14.3** Effect of adding sodium hydroxide solution to solutions containing various metal ions

Added dropwise	To excess	Likely cation
White precipitate	Precipitate is soluble in excess, giving a colourless solution	Al^{3+}, Zn^{2+}
White precipitate	Precipitate is insoluble in excess	Ca^{2+}
Green precipitate	Precipitate is insoluble in excess	Cr^{3+}
Light blue precipitate	Precipitate is insoluble in excess	Cu^{2+}
Green precipitate	Precipitate is insoluble in excess, turns brown near the surface on standing (Figure 14.28)	Fe^{2+}
Red-brown precipitate	Precipitate is insoluble in excess (Figure 14.28)	Fe^{3+}

In the case of ammonium salts containing the ammonium ion, NH_4^+, ammonia gas is produced on warming. The ammonium cation does not form an insoluble hydroxide. However, it forms ammonia and water upon heating.

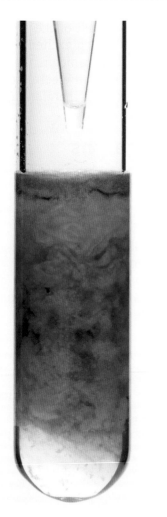

▲ Figure 14.28 Addition of sodium hydroxide will show the difference between Fe^{2+} and Fe^{3+} in aqueous solution

Effect of adding dilute ammonia solution

Ammonia gas dissolved in water is usually known as aqueous ammonia. The solution is only weakly alkaline, which results in a relatively low concentration of hydroxide ions. Aqueous ammonia can be used to identify salts of Al^{3+}, Ca^{2+}, Cr^{3+}, Cu^{2+}, Fe^{2+}, Fe^{3+} and Zn^{2+} ions. The colour of the precipitate or solution formed identifies the metal present (Table 14.4).

▼ Table 14.4 Effect of adding aqueous ammonia to solutions containing various metal ions

Added dropwise	To excess	Cation present
White precipitate	Precipitate is insoluble in excess	Al^{3+}
No precipitate or very slight white precipitate	No change	Ca^{2+}
Green precipitate	Precipitate is insoluble in excess	Cr^{3+}
Light blue precipitate	Precipitate is soluble in excess, giving a dark blue solution (Figure 14.29)	Cu^{2+}
Green precipitate	Precipitate is insoluble in excess, turns brown near the surface on standing	Fe^{2+}
Red-brown precipitate	Precipitate is insoluble in excess	Fe^{3+}
White precipitate	Precipitate is soluble in excess, giving a colourless solution	Zn^{2+}

▲ Figure 14.29 When aqueous ammonia is added to a solution containing Cu^{2+} ions the solution forms a gelatinous light blue precipitate. As more is added to excess, the precipitate dissolves, forming a dark blue clear solution

Tests for aqueous anions

Table 14.5 shows a variety of tests for aqueous anions.

▼ **Table 14.5** A variety of tests for aqueous anions

Anion	Test	Test result
Carbonate (CO_3^{2-})	Add dilute acid	Effervescence is seen as carbon dioxide produced
Chloride (Cl^-) [in solution]	Acidify with dilute nitric acid, then add aqueous silver nitrate	A white precipitate is produced
Bromide (Br^-) [in solution]	Acidify with dilute nitric acid, then add aqueous silver nitrate	A cream precipitate is produced
Iodide (I^-) [in solution]	Acidify with dilute nitric acid, then add aqueous silver nitrate	A yellow precipitate is produced
Nitrate (NO_3^-) [in solution]	Add aqueous sodium hydroxide, then aluminium foil; warm carefully	Ammonia gas is produced
Sulfate (SO_4^{2-}) [in solution]	Acidify, then add aqueous barium nitrate or barium chloride	A white precipitate is produced (Figure 14.30)
Sulfite (SO_3^{2-})	Add a small volume of acidified aqueous potassium manganite(VIII)	The acidified aqueous potassium manganate(VII) changes colour from purple to colourless

▲ **Figure 14.30** The test for sulfate ions

Tests for gases

Table 14.6 shows the common gases which may be produced during qualitative analysis and tests which can be used to identify them. These tests are used in conjunction with the tests shown above.

▼ **Table 14.6** Tests for gases

Gas	Colour (odour)	Effect of moist indicator paper	Test
Hydrogen (H_2)	Colourless (odourless)	No effect – neutral	'Pops' in the presence of a lighted splint
Oxygen (O_2)	Colourless (odourless)	No effect – neutral	Relights a glowing splint (Figure 14.31)
Carbon dioxide (CO_2)	Colourless (odourless)	Pink – weakly acidic	Turns limewater a cloudy white
Ammonia (NH_3)	Colourless (very pungent smell)	Blue – alkaline	Turns damp red litmus blue
Sulfur dioxide (SO_2)	Colourless (very choking smell)	Red – acidic	Turns acidified potassium dichromate(VI) from orange to green Turns acidified potassium manganate(VII) from purple to colourless
Chlorine (Cl_2)	Yellow-green (very choking smell)	Bleaches moist indicator paper after it initially turns pale pink	Bleaches damp litmus paper
Water (H_2O)	Colourless (odourless)	No effect – neutral	Turns blue cobalt chloride paper pink Turns anhydrous copper(II) sulfate from white to blue

▲ **Figure 14.31** Testing for oxygen gas

Practical skills

Testing a compound

Safety

● Eye protection must be worn.

A student was given small quantities of a compound labelled A. After putting on their eye protection the student carried out a variety of tests to help them identify compound A. Table 14.7 shows the tests the student did on compound A and the observations made.

Complete the table below by giving the possible conclusions to tests (a), (b), (c) and (d). Suggest both the test and observation that lead to the conclusion in test (d).

1. What is the name and formula for compound A?
2. It was suggested by the teacher that the final test (d) should be carried out in a fume cupboard. Explain why this procedure should be carried out in a fume cupboard.

▼ **Table 14.7** Tests on compound A and observations made

Test	Observation	Conclusion
(a) Compound A was dissolved in water and the solution produced divided into three parts for the tests (b), (c) and (d).	When dissolved in water a colourless solution was produced.	
(b) (i) To the first sample, a small amount of aqueous sodium hydroxide was added. (ii) An excess of aqueous sodium hydroxide was added to the mixture from (i).	(i) White precipitate was produced. (ii) The white precipitate dissolved in excess sodium hydroxide.	
(c) (i) To the second part, a small amount of aqueous ammonia was added. (ii) An excess of aqueous ammonia was then added to the mixture from (i).	(i) A white precipitate was formed. (ii) The white precipitate was insoluble in excess.	
(d) Aqueous sodium hydroxide solution was added to some of the solution of A followed by some aluminium foil. The mixture was then warmed carefully.	A gas was produced which turned moist litmus paper blue.	

▶ Test yourself

4. What will be the colour of the precipitate formed with silver nitrate for the following:
   a chloride, Cl⁻
   b bromide, Br⁻
   c iodide, I⁻
   Choose from yellow, white, green, cream, blue.
5. Match the gas shown in the table to the correct test for that gas and the positive result.

Gas	Test result
1 Oxygen	a Damp indicator paper turns white (bleached)
2 Carbon dioxide	b Burning splint produces a 'pop' sound when plunged into a test tube of gas
3 Hydrogen	c Damp litmus paper turns blue
4 Ammonia	d Limewater turns milky white
5 Chlorine	e Relights a glowing splint

Revision checklist

After studying Chapter 14 you should be able to:
- ✔ Name appropriate apparatus for the measurement of time, temperature, mass and volume.
- ✔ Suggest advantages and disadvantages of experimental methods and apparatus.
- ✔ Use the terms 'solvent', 'solute', 'solution', 'saturated solution', 'residue' and 'filtrate'.
- ✔ Describe an acid–base titration and know how to identify the end-point of a titration.
- ✔ Describe how paper chromatography is used to separate mixtures of soluble substances.
- ✔ Describe the use of locating agents when separating mixtures.
- ✔ Interpret simple chromatograms.
- ✔ State and use the equation:

$$R_f = \frac{\text{distance travelled by solute}}{\text{distance travelled by solvent}}$$

- ✔ Describe and explain methods of separation and purification.
- ✔ Suggest suitable separation and purification techniques.
- ✔ Identify substances and assess their purity using melting point and boiling point information.
- ✔ Describe tests using aqueous sodium hydroxide and aqueous ammonia to identify aqueous cations mentioned in this chapter in Tables 14.3 and 14.4.
- ✔ Describe the use of a flame test to identify the cations mentioned in this chapter in Table 14.2.
- ✔ Describe tests to identify the anions mentioned in this chapter in Table 14.5.
- ✔ Describe tests to identify the gases mentioned in this chapter in Table 14.6.

Exam-style questions

1 a Which of the following:
   cm³, kilograms, hours, degrees Celsius, dm³,
   minutes, seconds, grams
   are units of:
   i   time   [1]
   ii temperature   [1]
   iii volume   [1]
   iv mass?   [1]
  b In experiments you will require the use
   of accurate measuring instruments. State
   the precision you would expect from the
   following measuring instruments:
   i   a stopwatch   [1]
   ii a thermometer   [1]
   iii an electronic balance   [1]
   iv a burette.   [1]

2 State the method which is most suitable for
  separating the following:
  a oxygen from liquid air   [1]
  b red blood cells from plasma   [1]
  c petrol and kerosene from crude oil   [1]
  d coffee grains from coffee solution   [1]
  e pieces of steel from engine oil   [1]
  f amino acids from fruit juice solution   [1]
  g ethanol and water.   [1]

3 Three solid dyes are to be separated. One is
  yellow, one is red and one blue. The following
  facts are known about the dyes.
  a The blue and yellow dyes are soluble in
     cold water.
  b The red dye is insoluble.
  c When chalk is added to the solution of blue
     and yellow dyes, the mixture stirred and then
     filtered, the filtrate is blue and the residue is
     yellow.
  d When the residue is mixed with alcohol and
     the mixture stirred and filtered, the filtrate is
     yellow and the residue is white.
  Describe an experiment to obtain solid samples of
  each of the dyes from a mixture of all three.   [10]

4 Forensic scientists recovered several samples
  from a crime scene. Chromatography was carried
  out on these samples. The scientists placed
  the samples onto chromatography paper. The
  chromatogram produced was sprayed with a
  locating agent.
  The R_f *values* were then measured for each
  of the samples taken.

a What do you understand by the term:
   i   locating agent?   [2]
   ii R_f value?   [2]
  b The samples were thought to be amino acids.
   Use your research skills to find out the normal
   locating agent used for amino acids.   [1]
  c The table below shows R_f values for several
   amino acids thought to be at the crime
   scene and for three samples.

Amino acid	R_f value
Alanine	0.65
Glycine	0.25
Threonine	0.57
Proline	0.60
Lysine	0.16
Methionine	0.50
Sample 1	0.64
Sample 2	0.18
Sample 3	0.24

   i   Identify the samples taken from the crime
       scene using the data from the table.   [3]
   ii Explain how you made your choices
       in part i.   [1]

5 For each of the following pairs of substances,
  describe a chemical test you would carry out
  to distinguish between them.
  a potassium sulfate and potassium sulfite   [3]
  b ammonium chloride and aluminium chloride   [3]
  c zinc nitrate and calcium nitrate   [3]
  d sodium chloride and sodium iodide   [3]
  e iron(II) sulfate and copper(II) sulfate   [3]

6 Sodium carbonate hydrate contains water of
  crystallisation. When it is heated strongly,
  it gives off the water of crystallisation, which
  can be collected.
  a The substance left behind is anhydrous
     sodium carbonate. Describe a chemical test
     to show that this substance contains sodium
     (cation) and carbonate (anion).   [2]
  b Describe two chemical tests to show that the
     colourless liquid produced and collected is
     indeed water.   [4]
  c Describe another test to show that the
     colourless liquid given off in this experiment
     is pure water.   [1]

Theory past paper questions

1 Two isotopes of sulfur are $^{32}_{16}S$ and $^{33}_{16}S$.
   a What is meant by the term *isotopes*? [1]
   b Complete the table for $^{33}_{16}S$. [3]

number of neutrons	
number of protons	
electronic configuration	

   c Sulfur forms simple molecules which have a relative molecular mass of 256.
   Suggest the formula of a sulfur molecule. [1]
   d Sulfur has a low melting point and does not conduct electricity.
      i Explain why sulfur has a low melting point. [1]
      ii Explain why sulfur does not conduct electricity. [1]
   e Sulfur reacts with potassium to form potassium sulfide.
   Write the formula and the electronic configuration of the positive ion and of the negative ion in potassium sulfide.
   positive ion
   formula _____
   electronic configuration _____

   negative ion
   formula _____
   electronic configuration _____
   [2]
   f Sulfur reacts with hydrogen to form hydrogen sulfide, H_2S.
   Draw the 'dot-and-cross' diagram to show the bonding in a molecule of hydrogen sulfide.
   Only draw the outer shell electrons. [2]
   g Hydrogen sulfide reacts with sulfur dioxide to form sulfur and water.
   Write the equation for this reaction. [1]
   [Total: 12]

(Cambridge O Level Chemistry (5070) Paper 21, Q A4, May/June 2015)

2 An aqueous solution of ammonium nitrite, NH_4NO_2, decomposes when heated gently.
$$NH_4NO_2(aq) \rightarrow N_2(g) + 2H_2O(l)$$

   a Describe how you could show that aqueous ammonium nitrite contains ammonium ions. [2]
   b A sample of $25.0\,cm^3$ of $0.500\,mol/dm^3$ aqueous ammonium nitrite is heated.
   Calculate the volume of nitrogen formed, measured at room temperature and pressure.
   volume of nitrogen = _____ [3]
   c Ammonium nitrate, NH_4NO_3, decomposes when heated, in a similar way to ammonium nitrite.
   Suggest the formulae of the two products made in this reaction. [1]
   d Describe how a pure sample of aqueous ammonium nitrate can be prepared from dilute nitric acid. [4]
   [Total: 10]

(Cambridge O Level Chemistry (5070) Paper 21, Q B6, May/June 2015)

3 Molybdenum is a transition element.
   It is used to make steel that is extremely hard.
   Molybdenum can be manufactured by heating together molybdenum(VI) oxide, MoO_3, and aluminium.
   a Construct the equation for this reaction. [1]
   b Explain why this reaction involves both oxidation and reduction. [1]
   c What mass of molybdenum can be made from $125\,g$ of molybdenum(VI) oxide?
   [A_r: Mo, 96] [3]
   d Which metal is the less reactive, aluminium or molybdenum?
   Explain your answer. [1]
   e Molybdenum has a melting point of $2623\,°C$.
      i Describe metallic bonding, with the aid of a labelled diagram. [2]
      ii Suggest why molybdenum has a much higher melting point than aluminium. [2]
   [Total: 10]

(Cambridge O Level Chemistry (5070) Paper 21, Q B7, May/June 2015)

4 Large quantities of poly(chloroethene) are manufactured annually.
   The flow chart shows the steps involved in the manufacture of poly(chloroethene).

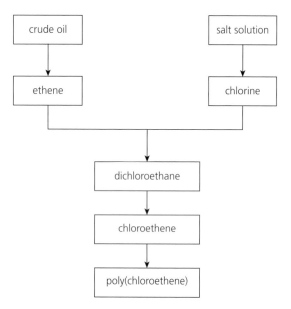

a Name the **two** processes used to
   manufacture ethene from crude oil. [2]
b The salt solution is electrolysed using a
   carbon anode (positive electrode).
   Write the equation for the reaction
   occurring at the anode. [1]
c Draw the structure, showing all the atoms
   and all the bonds, of the dichloroethane. [1]
d When dichloroethane, $C_2H_4Cl_2$, is heated
   strongly chloroethene, C_2H_3Cl, is formed.
   Name the other product of this reaction. [1]
e The structure of chloroethene is shown.

$$\underset{H}{\overset{H}{\diagdown}} C = C \underset{Cl}{\overset{H}{\diagup}}$$

   Draw part of the structure of
   poly(chloroethene). [2]
f A factory uses 2250 tonnes of chloroethene to
   make poly(chloroethene).
   i Deduce the maximum mass of
      poly(chloroethene) the factory could make.

      maximum mass = _____ tonnes [1]
   ii The actual yield of poly(chloroethene) is
       2175 tonnes.
       Calculate the percentage yield.
       percentage yield = _____ % [2]
       [Total: 10]

(Cambridge O Level Chemistry (5070) Paper 21, Q B8,
May/June 2015)

5 Alkanes are a homologous series of saturated
   hydrocarbons.
   The table shows information about some alkanes.

alkane	molecular formula	melting point/°C	boiling point/°C
ethane	C_2H_6	−183	−89
butane	C_4H_{10}	−138	0
hexane	C_6H_{14}	−95	69
decane	$C_{10}H_{22}$	−30	174
dodecane	$C_{12}H_{26}$	−10	216

a Dodecane is a liquid at 25 °C.
   How can you make this deduction from
   the data in the table? [2]
b Butane melts at −138 °C.
   Use the kinetic particle theory to explain
   what happens when butane melts. [2]
c A sample of ethane gas at 0 °C is at a
   pressure of 1 atmosphere.
   The pressure is increased but the
   temperature is maintained at 0 °C.
   Describe and explain, in terms of the
   kinetic particle theory, what happens to
   the volume of the gas. [2]
d Suggest a method of separating a mixture
   of hexane, decane and dodecane.
   Explain your answer. [1]
e Draw the structure, showing all the atoms
   and all the bonds, for two isomers with the
   molecular formula C_4H_{10}. [2]
f The structure of hexane is shown.

$$H-\overset{\overset{\displaystyle H}{|}}{C}-\overset{\overset{\displaystyle H}{|}}{C}-\overset{\overset{\displaystyle H}{|}}{C}-\overset{\overset{\displaystyle H}{|}}{C}-\overset{\overset{\displaystyle H}{|}}{C}-\overset{\overset{\displaystyle H}{|}}{C}-H$$

   Draw the structure, showing all the atoms
   and all the bonds, of an organic product of
   the reaction of hexane with chlorine. [1]
   [Total: 10]

(Cambridge O Level Chemistry (5070) Paper 21, Q B9,
May/June 2015)

6 Hydrogen sulfide, H_2S, has a simple molecular
   structure. It is soluble in water.
   a Suggest **one** other physical property of
      hydrogen sulfide. [1]
   b Aqueous hydrogen sulfide is a weak acid.
      i Write an equation to show the
         dissociation of hydrogen sulfide. [1]

ii Why is aqueous hydrogen sulfide described as a weak acid? [1]

c Aqueous hydrogen sulfide reacts with aqueous potassium hydroxide.

$$H_2S(aq) + 2KOH(aq) \rightarrow K_2S(aq) + 2H_2O(l)$$

What is the minimum volume, in cm³, of $0.150\,mol/dm^3$ KOH required to completely react with a solution containing $0.170\,g$ of H_2S? [3]

d Magnesium reacts with sulfur to make the ionic compound magnesium sulfide.
   i Predict **two** physical properties of magnesium sulfide. [2]
   ii Explain, in terms of electrons, how a magnesium atom reacts with a sulfur atom to make a magnesium ion and a sulfide ion. [2]
   [Total: 10]

(Cambridge O Level Chemistry (5070) Paper 21, Q A2, May/June 2016)

7 Ammonia is manufactured by the reaction between hydrogen and nitrogen in the Haber process.
   a State the conditions used in the Haber process.
   temperature _____
   pressure _____
   catalyst _____ [2]
   b Describe and explain the effect of increasing the pressure on the **rate** of this reaction. [2]
   c Explain how a catalyst speeds up the rate of a chemical reaction. [1]
   d Ammonia is used to make fertilisers. The table gives some information about two fertilisers made from ammonia.

fertiliser	formula	relative formula mass (M_r)
ammonium nitrate	NH_4NO_3	80
urea	$(NH_2)_2CO$	60

Use the data in the table to show that urea contains a greater percentage by mass of nitrogen than ammonium nitrate. [2]
   [Total: 7]

(Cambridge O Level Chemistry (5070) Paper 21, Q A4, May/June 2016)

8 Silver nitrate has the formula $AgNO_3$.
   a Describe how a pure sample of silver nitrate crystals can be prepared from solid silver oxide, which is insoluble in water. [4]
   b Aqueous zinc chloride is added to a sample of acidified aqueous silver nitrate.
   i Describe what you would observe. [1]
   ii Construct the ionic equation, with state symbols, for the reaction that occurs. [2]
   c Aqueous silver nitrate is electrolysed using graphite electrodes.
   i Identify the product formed at the cathode (negative electrode). [1]
   ii Oxygen and water are formed at the anode (positive electrode). Construct the equation for the reaction at the anode. [1]
   d Silver nitrate decomposes on heating to form Ag_2O, NO_2 and O_2. Construct the equation for this reaction. [1]
   [Total: 10]

(Cambridge O Level Chemistry (5070) Paper 21, Q B7, May/June 2016)

9 Carbon reacts with steam in a reversible reaction.
   $$C(s) + H_2O(g) \rightleftharpoons H_2(g) + CO(g) \quad \Delta H = +131\ kJ/mol$$
   The reaction reaches an equilibrium if carried out in a closed container.
   a Explain, in terms of bond breaking and bond forming, why this reaction is endothermic. [2]
   b When one mole of hydrogen, H_2, is formed, 131 kJ of energy is absorbed. Calculate the amount of energy absorbed when 240 dm³ of hydrogen, measured at room temperature and pressure, is formed. [2]
   c Predict, with a reason, how the **position of equilibrium** of this reaction changes as the
   i pressure is increased at constant temperature, [2]
   ii temperature is increased at constant pressure. [2]
   d The reaction between carbon and steam is a possible source of hydrogen.
   i Suggest one disadvantage of using this reaction as a source of hydrogen. [1]
   ii Another source of hydrogen is the cracking of hydrocarbons from crude oil.

Give one advantage of manufacturing hydrogen from the reaction of carbon with steam rather than from crude oil.   [1]
[Total: 10]

(Cambridge O Level Chemistry (5070) Paper 21, Q B9, May/June 2016)

10 a Atoms and ions contain three types of sub-atomic particle.
Complete the table about these sub-atomic particles.   [3]

sub-atomic particle	relative charge	relative mass
electron		
neutron		1
proton	+1	

b The table shows some information about six particles.

particle	number of protons in particle	number of neutrons in particle	number of electrons in particle
A	37	48	37
B	53	74	54
C	92	143	92
D	92	143	89
E	92	146	92
F	94	150	92

i   What is the nucleon number for particle **A**?   [1]
ii   Explain why particle **B** is a negative ion.   [1]
iii Which two **atoms** are isotopes of the same element?
Explain your answer.   [2]
[Total: 7]

(Cambridge O Level Chemistry (5070) Paper 21, Q A2, May/June 2017)

11 Acids are neutralised by alkalis.
a Write the ionic equation for the reaction between an acid and an alkali.   [1]
b Sodium sulfate is a soluble salt that can be prepared using a titration method.
i Name a sodium compound and the acid that can be used to make sodium sulfate by this method.   [1]

ii Describe how the titration method is used to prepare a colourless solution of sodium sulfate.   [3]
iii Describe how a sample of pure sodium sulfate crystals can be made from aqueous sodium sulfate.   [2]
c Aqueous sodium sulfate can be used to prepare barium sulfate.

$$Ba^{2+}(aq) + SO_4^{2-}(aq) \rightarrow BaSO_4(s)$$

In an experiment, 20.0 cm³ of 0.550 mol/dm³ of barium nitrate was added to excess aqueous sodium sulfate.
i   Calculate the maximum mass of barium sulfate that could be made.
[The relative formula mass of $BaSO_4$ is 233.]
maximum mass of barium sulfate

= _____ g   [2]
ii   A mass of 1.92 g of dry barium sulfate was obtained. Calculate the percentage yield of barium sulfate.
percentage yield of barium sulfate

= _____ %   [1]
[Total: 10]

(Cambridge O Level Chemistry (5070) Paper 21, Q A3, May/June 2017)

12 Calcium chloride, $CaCl_2$, is an ionic compound.
a State the electronic configuration for each of the ions in calcium chloride.   [2]
b When **molten** calcium chloride is electrolysed, calcium and chlorine are formed.
Construct equations for the two electrode reactions.   [2]
c Predict the products of the electrolysis of concentrated **aqueous** calcium chloride.   [1]
d Explain, using ideas about structure and bonding, why calcium chloride has a high melting point.   [2]
[Total: 7]

(Cambridge O Level Chemistry (5070) Paper 21, Q A4, May/June 2017)

13 Ethanol and butanol are both alcohols.

ethanol   butanol

a Describe the manufacture of ethanol from ethene. [2]

b Ethanol is used as a fuel and as a constituent of alcoholic beverages.
   i State one **other** use of ethanol. [1]
   ii Construct an equation to show the **incomplete** combustion of ethanol. [2]

c Ethanol can be oxidised to form ethanoic acid.
   Name a reagent that can be used for this oxidation. [1]

d Draw the structure of an alcohol that is an isomer of butanol.
   Show all of the atoms and all of the bonds. [1]

e Butanol can be converted into an alkene by loss of a molecule of water.
   Draw the structure of the alkene formed.
   Show all of the atoms and all of the bonds. [1]

f Butene can be polymerised to give poly(butene).
   i What type of polymerisation occurs? [1]
   ii Poly(butene) is non-biodegradable.
   What does the term *non-biodegradable* mean? [1]

[Total: 10]

(Cambridge O Level Chemistry (5070) Paper 21, Q A5, May/June 2017)

14 Copper reacts with concentrated nitric acid.
$Cu(s) + 4HNO_3(aq) \rightarrow Cu(NO_3)_2(aq) + 2NO_2(g) + 2H_2O(l)$

a Suggest what you would observe when copper reacts with concentrated nitric acid. [1]

b **i** Suggest the name of the salt of formula $Cu(NO_3)_2$. [1]
   ii Copper is oxidised when it reacts with concentrated nitric acid.
   Use the equation to explain that copper has been oxidised. [1]

c An excess of copper is added to $25.0\,cm^3$ of $16.0\,mol/dm^3$ HNO_3.
   Use this information, together with the equation above, to calculate the volume of NO_2 formed.
   The gas volume is measured at room temperature and pressure. [3]

d When heated, $Cu(NO_3)_2$ decomposes to form CuO, NO_2 and O_2.
   Construct the equation for this reaction. [1]

e To a sample of $Cu(NO_3)_2(aq)$, a student adds aqueous ammonia drop by drop until it is in excess.
   i Describe what is observed. [2]
   ii The student repeats the experiment but adds aqueous sodium hydroxide instead of aqueous ammonia.
   Describe what is observed. [1]

[Total: 10]

(Cambridge O Level Chemistry (5070) Paper 21, Q B7, May/June 2017)

15 The elements in Group VII of the Periodic Table are called the halogens.

a Explain why the elements in Group VII have similar chemical properties. [1]

b A redox reaction happens when chlorine gas is bubbled through aqueous potassium iodide.
   $Cl_2(g) + 2I^-(aq) \rightarrow I_2(aq) + 2Cl^-(aq)$
   i Describe what is observed during this reaction. [1]
   ii Use the equation to explain that oxidation takes place in this reaction. [1]
   iii Use the equation to explain that reduction takes place in this reaction. [1]

c Describe the chemical test for chlorine gas.
   test _____
   observation _____ [2]

d Chlorine reacts with iron to form iron(III) chloride in a closed container.
   i The pressure of chlorine is increased.
   Describe and explain what happens to the rate of this reaction. [2]
   ii Iron(III) chloride can act as a catalyst for some reactions.
   Explain how a catalyst increases the rate of a reaction. [2]

[Total: 10]

(Cambridge O Level Chemistry (5070) Paper 21, Q 2, May/June 2018)

16 Barium chloride is a soluble salt and barium sulfate is an insoluble salt.

a Barium sulfate can be prepared by the reaction between aqueous barium chloride and dilute sulfuric acid.
   i Describe the preparation of a pure, dry sample of barium sulfate from aqueous barium chloride and dilute sulfuric acid. [3]
   ii Write the ionic equation, including state symbols, for this reaction. [2]

b Barium chloride can be prepared by reacting barium carbonate with dilute hydrochloric acid.

$BaCO_3 + 2HCl \rightarrow BaCl_2 + H_2O + CO_2$

Excess barium carbonate is reacted with $40.0\,cm^3$ of $1.50\,mol/dm^3$ hydrochloric acid.

After purification the percentage yield of barium chloride was 75.0%.

Calculate the mass of barium chloride prepared.

Give your answer to **three** significant figures.

[M_r: $BaCl_2$, 208] [3]

c A barium ion has the formula $^{138}_{56}Ba^{2+}$.

Complete the table about this ion. [3]

subatomic particles	number of subatomic particles
electrons	
neutrons	
protons	

[Total: 11]

(Cambridge O Level Chemistry (5070) Paper 21, Q 3, May/June 2018)

17 A scientist investigates the reaction of iron with steam in a closed system.

A dynamic equilibrium mixture is established.

$3Fe(s) + 4H_2O(g) \rightleftharpoons Fe_3O_4(s) + 4H_2(g)$

a Explain why the concentrations of steam and of hydrogen do not change once the dynamic equilibrium mixture has been established. [1]

b The pressure of the equilibrium mixture is increased.

The temperature of the closed system is kept constant.

Predict and explain what will happen, if anything, to the composition of the equilibrium mixture. [2]

c The temperature of the equilibrium mixture is increased.

The pressure within the closed system is kept constant.

i The position of equilibrium shifts to the left hand side.

What conclusion can be made about the enthalpy change of the reaction? [1]

ii Describe and explain what happens to the rate of reaction. [1]

d Dilute sulfuric acid reacts with Fe_3O_4 to form three compounds, **A**, **B** and **C**.

- **A** is iron(II) sulfate.
- **B** is iron(III) sulfate.
- **C** is a colourless liquid.

i Name compound **C**. [1]

ii Construct the equation for this reaction. [2]

iii Describe a chemical test for iron(III) ions.

test _____

observation _____ [2]

[Total: 10]

(Cambridge O Level Chemistry (5070) Paper 21, Q 7, May/June 2018)

18 Choose from the particles shown to answer the questions.

$^{79}_{35}Br$ $^{39}_{20}Ca$

$^{35}_{17}Cl$ $^{37}_{17}Cl^-$

$^{37}_{17}Cl$ $^{64}_{29}Cu$

$^{23}_{11}Na$ $^{20}_{10}Ne$

$^{17}_{8}O$ $^{18}_{8}O^{2-}$

Each particle can be used once, more than once or not at all.

a Which particle has only 20 protons in its nucleus? [1]

b Which particle has a nucleon number of 35? [1]

c Which particle has an electronic structure of 2.8.8? [1]

d Which particle is an atom with only 10 neutrons in its nucleus? [1]

e Which particle is an atom of a transition element? [1]

[Total: 5]

(Cambridge O Level Chemistry (5070) Paper 21, Q 1, May/June 2019)

19 The table shows some of the properties of the elements in Group II of the Periodic Table.

element	proton (atomic) number	atomic radius/nm	melting point/°C
Be	4	0.089	1280
Mg	12	0.136	650
Ca	20	0.174	850
Sr	38	0.191	768
Ba	56	0.198	714
Ra	88		

a Explain why the elements in Group II have similar chemical properties. [1]

b Explain why it is easier to predict the atomic radius of radium, Ra, than the melting point of radium. [1]

c Magnesium chloride contains Mg^{2+} and Cl^- ions.
  i Write the electronic configuration for a magnesium ion. [1]
  ii Magnesium is produced by the electrolysis of molten magnesium chloride.
  Construct equations for the reactions taking place at the:
  negative electrode
  positive electrode. [2]

d Magnesium reacts with aqueous copper(II) sulfate in a redox reaction.

$$Mg(s) + Cu^{2+}(aq) \rightarrow Mg^{2+}(aq) + Cu(s)$$

Which particle is reduced?
Explain your answer. [1]

e Magnesium reacts with steam.
Name the products of this reaction. [1]

f Calcium reacts with cold water.
Write the equation for this reaction. [1]

g Magnesium chloride is a soluble salt.
Describe how a pure sample of magnesium chloride crystals can be made from magnesium. [4]
[Total: 12]

(Cambridge O Level Chemistry (5070) Paper 21, Q 2, May/June 2019)

20 Acid **U** is a compound containing carbon, hydrogen and oxygen.
a A 6.30 g sample of **U** contains 1.68 g of carbon and 0.14 g of hydrogen.
Calculate the empirical formula of **U**. [3]

b A 0.086 g sample of **U** is completely neutralised by 12.7 cm³ of 0.150 mol/dm³ KOH.
One mole of **U** reacts with two moles of KOH.
Calculate the relative formula mass of **U**. [3]

c What is the molecular formula of **U**? [1]
[Total: 7]

(Cambridge O Level Chemistry (5070) Paper 21, Q 5, May/June 2019)

21 A scientist heats a sample of phosphorus(V) chloride in a closed container.
A dynamic equilibrium is established.

$$PCl_5(g) \rightleftharpoons PCl_3(g) + Cl_2(g)$$

a Describe what is meant by the term *dynamic equilibrium*. [2]

b The pressure of the equilibrium mixture is increased.
The temperature of the equilibrium mixture is kept constant.
Predict and explain what will happen, if anything, to the **composition** of the equilibrium mixture. [2]

c The temperature of the equilibrium mixture is increased.
The pressure of the equilibrium mixture is kept constant.
  i Suggest why the position of equilibrium moves to the right. [1]
  ii Explain why the rate of the reaction increases. [2]

d Draw the 'dot-and-cross' diagram for a molecule of PCl_3.
Only include the outer shell electrons. [2]

e PCl_5 reacts with water to form hydrogen chloride and phosphoric acid, H_3PO_4.
Construct an equation for this reaction. [1]
[Total: 10]

(Cambridge O Level Chemistry (5070) Paper 21, Q 8, May/June 2019)

22 Cyclobutane and butene are both hydrocarbons.

cyclobutane   butene

a What is meant by the term *hydrocarbon*? [1]

b Explain why cyclobutane and butene are isomers. [1]

c Cyclobutane is saturated and butene is unsaturated.
Describe a chemical test that can distinguish cyclobutane from butene.
test _____
result for cyclobutane _____
result for butene _____. [3]

d Calculate the percentage by mass of carbon in butene. [2]

e Ethene can be converted into ethanoic acid in a two-step process.

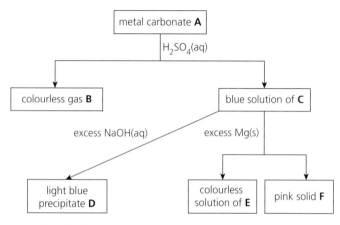

i Identify compound **X**. [1]
ii Identify the reagent used in step **1**. [1]
iii Identify the reagent used in step **2**. [1]
[Total: 10]

(Cambridge O Level Chemistry (5070) Paper 21, Q 10, May/June 2019)

23 Aluminium is manufactured by the electrolysis of aluminium oxide dissolved in molten cryolite.

a Give the equations for the reactions that occur at the electrodes during this electrolysis.

positive electrode _____

negative electrode _____ [2]

b Aluminium is a useful metal as it does not corrode in moist air.
Explain why aluminium does not corrode in moist air. [2]

c Underground iron pipes rust easily. This can be prevented by attaching a piece of magnesium to the pipe.
Explain this form of rust prevention. [2]

d Aluminium sulfate is a soluble salt.
Describe how a sample of aluminium sulfate crystals can be prepared from aluminium oxide. [4]
[Total: 10]

(Cambridge O Level Chemistry (5070) Paper 21, Q A4, May/June 2014)

24 The flow chart shows some reactions of the compounds of a metal.

Identify, by name, each of the substances.

A _____

B _____

C _____

D _____

E _____

F _____

[Total: 6]

(Cambridge O Level Chemistry (5070) Paper 21, Q A6, May/June 2014)

25 When carbon dioxide reacts with hydrogen in a sealed container, an equilibrium mixture is obtained.

$$CO_2(g) + 4H_2(g) \rightleftharpoons CH_4(g) + 2H_2O(g) \ \Delta H = -205 \text{ kJ/mo}$$

This reaction is exothermic.

a Describe and explain what happens to the rate of the forward reaction when the pressure is increased. The temperature remains constant. [2]

b Describe and explain what happens to the position of equilibrium when the temperature is increased. The pressure remains constant. [2]

c In an experiment, 220 g of carbon dioxide and an excess of hydrogen are reacted in a sealed container until an equilibrium is established.

A mass of 46 g of methane is produced.

i Calculate the mass of methane that should have been made if the percentage yield was 100%. [2]

ii Calculate the percentage yield of methane in this experiment. [1]

d The experiment with 220 g of carbon dioxide and an excess of hydrogen is repeated but this time a catalyst is added.

i State what happens, if anything, to the position of equilibrium compared with the non-catalysed reaction. [1]

ii Describe and explain what happens to the rate of reaction compared with the non-catalysed reaction. [2]

[Total: 10]

(Cambridge O Level Chemistry (5070) Paper 21, Q B9, May/June 2014)

Alternative to Practical past paper questions

1 **H** is a compound which contains three ions. Complete the table by adding the conclusion for **a**, the observations for **b i**, **ii** and **iii**, and both the test and observation for **c**. Any gas evolved should be tested and named.

test	observations	conclusions
a **H** is dissolved in water and the resulting solution divided into two parts for use in tests **b** and **c**.	A coloured solution is formed.	
b i To the first part, aqueous sodium hydroxide is added until a change is seen.		**H** contains Fe^{2+} ions.
ii An excess of aqueous sodium hydroxide is added to the mixture from **i**.		**H** contains Fe^{2+} ions.
iii This mixture is warmed.		**H** contains NH_4^+ ions.
c		**H** contains SO_4^{2-} ions.

[Total: 8]

(Cambridge O Level Chemistry (5070) Paper 41, Q 8, May/June 2014)

2 **a** The general formula of a carboxylic acid is $C_nH_{2n+1}COOH$
  G is a carboxylic acid and has a relative formula mass of 88.
  Deduce the value of **n** in the formula and hence name compound **G**. [3]
  b When magnesium reacts with **G**, a gas is formed.
  Name the gas and give a test and observation to identify this gas.
  name _____
  test and observation _____. [2]
  c Ethyl ethanoate has the same molecular formula as **G**.
  i To which homologous series does ethyl ethanoate belong? [1]
  ii Give the structure of ethyl ethanoate. [1]

iii Name the two compounds which, under suitable conditions, react to form ethyl ethanoate. [2]
iv Compound **H** belongs to the same homologous series as ethyl ethanoate. **H** has the same molecular formula as ethyl ethanoate.
  Suggest the name and the structure of **H**.
  • name
  • structure [2]
  [Total: 11]

(Cambridge O Level Chemistry (5070) Paper 41, Q 2, May/June 2015)

3 **a** i What colour is litmus paper when dipped in hydrochloric acid? [1]
  ii Suggest two ways by which the pH of dilute hydrochloric acid can be measured. [2]
  iii Suggest a value for the pH of dilute hydrochloric acid. [1]
  b A student adds an equal volume of aqueous sodium carbonate separately to dilute ethanoic acid and dilute hydrochloric acid.
  i What does the student observe in both reactions? [1]
  ii Compare the rates of the two reactions and explain the difference. [2]
  c i A small amount of magnesium ribbon is added to a test-tube containing dilute hydrochloric acid. A gas is produced.
  Name the gas. Give a test and observation to identify the gas. [2]
  ii Construct the equation for the reaction. [1]
  [Total: 10]

(Cambridge O Level Chemistry (5070) Paper 41, Q 2, May/June 2016)

4 A student is given a sample of an organic acid, **V**, and asked to
  • determine its relative molecular mass,
  • suggest its molecular formula.
  A sample of the acid is placed in a previously weighed container and reweighed.
  mass of container + **V** = 8.38 g
    mass of container = 6.92 g

a Calculate the mass of **V** used in the experiment. [1]

The student transfers the sample to a beaker and adds 50.0 cm³ of 1.00 mol/dm³ sodium hydroxide, an excess. The contents are allowed to react and are then transferred to a volumetric flask. The solution is made up to 250 cm³ with distilled water. This is solution **W**.

25.0 cm³ of **W** is transferred into a conical flask. A few drops of thymolphthalein indicator are added to the conical flask. Thymolphthalein is colourless in acidic solution and blue in alkaline solution.

0.100 mol/dm³ hydrochloric acid is put into a burette and added to the solution in the conical flask until an end-point is reached.

b What is the colour of the solution in the conical flask
  • before the acid is added
  • at the end-point? [1]

c The student does three titrations. The diagrams show parts of the burette with the liquid levels at the beginning and end of each titration.

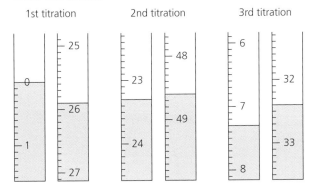

1st titration   2nd titration   3rd titration

Use the diagrams to complete the following table.

titration number	1	2	3
final burette reading/cm³			
initial burette reading/cm³			
volume of 0.100 mol/dm³ hydrochloric acid/cm³			
best titration results (✓)			

Summary
Tick (✓) the best titration results.

Using these results, the average volume of 0.100 mol/dm³ hydrochloric acid used is

_____ cm³. [4]

d Calculate the number of moles of hydrochloric acid in the average volume of 0.100 mol/dm³ hydrochloric acid from **(c)**. [1]

The equation for the reaction between hydrochloric acid and sodium hydroxide is shown.
$$HCl + NaOH \rightarrow NaCl + H_2O$$

e Using the equation and your answer from **(d)**, deduce the number of moles of sodium hydroxide in 25.0 cm³ of **W**. [1]

f Using your answer from **(e)**, calculate the number of moles of sodium hydroxide in 250 cm³ of **W**. [1]

g Calculate the number of moles of sodium hydroxide in 50 cm³ of 1.00 mol/dm³ sodium hydroxide. [1]

h By subtracting your answer in **(f)** from your answer in **(g)**, calculate the number of moles of sodium hydroxide that reacted with the original sample of the organic acid, **V**. [1]

i **One** mole of **V** reacts with **two** moles of sodium hydroxide.
Deduce the number of moles of **V** in the sample. [1]

j Using your answers from **(a)** and **(i)** calculate the relative molecular mass of the acid **V**. [1]

k The acid **V** contains two carboxylic acid groups and has the molecular formula
$$HO_2CC_xH_yCO_2H$$
where **x** and **y** are whole numbers.
Deduce the values of **x** and **y** in the molecular formula.
[A_r: H, 1; C, 12; O, 16] [2]

l Give the structure of the ester produced when **V** reacts with two molecules of ethanol under suitable conditions. [1]
[Total: 16]

(Cambridge O Level Chemistry (5070) Paper 41, Q 7, May/June 2016)

5 A student does an experiment to convert magnesium into magnesium oxide, MgO.
A 0.36 g sample of magnesium is heated strongly for several minutes using the apparatus shown.

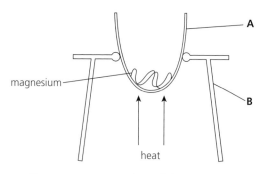

a Name apparatus **A** and **B**. [2]
Magnesium is converted into a white powder, MgO. The expected mass of MgO is 0.60 g. The student found that 0.55 g of MgO is produced in the experiment.

b Suggest one reason why the mass of MgO is lower than expected and suggest how the expected result may be achieved. [2]

c The student does a similar experiment using 0.36 g of zinc instead of 0.36 g of magnesium. Explain why he is wrong to expect that the mass of zinc oxide will also be 0.60 g.
[A_r: Mg, 24; Zn, 65] [2]

d Suggest a safety item that the student should use when doing this experiment. [1]
[Total: 7]

(Cambridge O Level Chemistry (5070) Paper 41, Q 1, May/June 2017)

6 Sand is insoluble in water; sodium chloride is soluble in water.
You are provided with a beaker containing 10.0 g of a mixture of sand and sodium chloride.
Suggest an experiment to determine the percentage, by mass, of sodium chloride in the mixture.
You should state
• the apparatus required,
• any measurements you need to make,
• how you would use your results to determine the percentage, by mass, of sodium chloride in the mixture. [5]
[Total: 5]

(Cambridge O Level Chemistry (5070) Paper 41, Q 2, May/June 2017)

7 A student is given compound **M** which contains a cation and an anion. He does the following tests to identify the two ions.

a A sample of **M** is dissolved in water. The solution is colourless.

Suggest what conclusion can be made. [1]

b To a test-tube containing 1 cm³ of aqueous **M**, a small volume of aqueous sodium hydroxide is added.
A white precipitate is produced. The precipitate is soluble in excess aqueous sodium hydroxide.
Suggest **two** cations which could be present in aqueous **M**. [2]

c What further test should the student do with aqueous **M** to identify which of the two cations suggested in **b** is present in **M**?
test _____
observations _____ [2]

d **M** is known to contain either chloride or iodide ions.
Suggest a test to identify which of the two anions is present in **M**.
test _____
observations _____ [3]
[Total: 8]

(Cambridge O Level Chemistry (5070) Paper 41, Q 5, May/June 2017)

8 A student does a series of titrations to determine the percentage of ethanoic acid in a sample of vinegar.
Diagrams of some of the apparatus used by the student are shown.

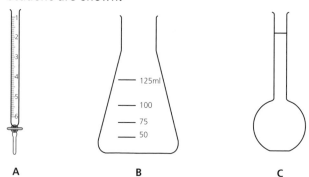

a Name the three pieces of apparatus. [3]

b The student measures 5.0 cm³ of the vinegar into apparatus **C** and makes it up to 250 cm³ with distilled water.
Apparatus **A** is filled with 0.0250 mol/dm³ sodium hydroxide.
For each titration, 25 cm³ of the diluted vinegar is transferred into apparatus **B**, using a measuring cylinder. A few drops of methyl orange indicator are added.

i The diagram shows parts of apparatus **A** with the liquid levels at the beginning and end of titration 4.
Record these values in the results table.
Calculate and record the volume of 0.0250 mol/dm³ sodium hydroxide used. [2]

titration 4

initial reading

— 0.0

— 1.0

final reading

— 18.0

— 19.0

titration number	1	2	3	4
final reading/cm³	19.0	36.4	19.1	
initial reading/cm³	0.0	18.4	0.4	
volume of 0.0250 mol/dm³ sodium hydroxide used / cm³		18.0		
best titration results (✓)				

ii Complete the results table by calculating the volume of 0.0250 mol/dm³ sodium hydroxide used for each of titrations 1 and 3. [1]
iii In the results table, tick (✓) the best titration results and use them to calculate the average titre. [1]
iv Suggest an improvement that the student can make to the method to make the results more accurate. Explain your answer. [2]
c A second student does another series of titrations using the same solutions. This student obtains an average titre of 18.4 cm³. The equation for the reaction that takes place during the titration is shown.

$$CH_3COOH + NaOH \rightarrow CH_3COONa + H_2O$$

i Calculate the number of moles of 0.0250 mol/dm³ sodium hydroxide used. [1]

ii Calculate the number of moles of ethanoic acid present in the 25 cm³ of diluted vinegar solution transferred into apparatus **B** for each titration. [1]
iii The diluted vinegar solution is made by making the original 5.0 cm³ of vinegar up to 250 cm³ with distilled water. Calculate the number of moles of ethanoic acid in the original 5.0 cm³ sample of vinegar. [1]
iv Calculate the concentration, in mol/dm³, of ethanoic acid in the original sample of vinegar. [1]

[Total: 13]

(Cambridge O Level Chemistry (5070) Paper 41, Q 1, May/June 2018)

9 A student investigates the order of reactivity of four metals by placing samples of each metal into aqueous solutions of the metal nitrates as shown.

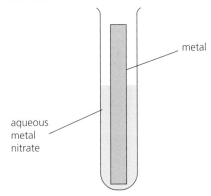

metal

aqueous metal nitrate

A results table is shown in which 'yes' indicates that a reaction took place and 'no' indicates there was no reaction.

metal	aqueous copper(II) nitrate	aqueous tin(II) nitrate	aqueous magnesium nitrate	aqueous zinc nitrate
copper		no	no	no
tin	yes		no	no
magnesium	yes	yes		yes
zinc	yes	yes	no	

a Describe what the student observes when zinc is placed into aqueous copper(II) nitrate. [2]
b i Use the experimental results to deduce the order of reactivity of these four metals, from most reactive to least reactive. [1]

   ii Explain how the results support this order of reactivity. [2]

  c The order of reactivity of these four metals can also be found by measuring the temperature change when each metal reacts with dilute hydrochloric acid.
Describe how you would do this experiment. Your description should include:

- the measurements you need to take
- the variables you need to keep constant
- an explanation of how the order of reactivity can be deduced from your results. [5]

[Total: 10]

(Cambridge O Level Chemistry (5070) Paper 41, Q 4, May/June 2018)

10 In the presence of a catalyst, hydrogen peroxide, H_2O_2, decomposes into water and oxygen.
A student uses the apparatus shown to investigate the rate of decomposition of samples of hydrogen peroxide at two different temperatures.

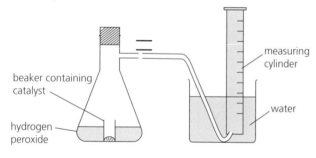

The experiment starts when the flask is tipped so that the catalyst comes into contact with the hydrogen peroxide.

  a The oxygen gas is collected in the measuring cylinder.

    i What property of oxygen gas allows it to be collected by this method? [1]

    ii Name an alternative piece of apparatus that could be used to collect and measure the volume of oxygen gas. [1]

    iii Give a test and observation to identify oxygen.

- test
- observation. [2]

  b The results obtained for the first experiment, at 25 °C, are shown.

time/min	0	2	4	6	8	10	12	14
volume of oxygen gas/ cm³	0	40	50	91	97	99	100	100

    i Plot the results on the grid. [2]

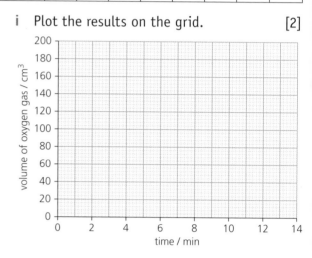

    ii Draw a circle around the anomalous point on the graph. [1]

    iii Use the points to draw a curve of best fit. [1]

    iv The student repeats the experiment at 50 °C. All other variables are kept constant. Draw a second curve on the grid to represent the results that are obtained at this higher temperature.
Explain your answer. [4]

[Total: 12]

(Cambridge O Level Chemistry (5070) Paper 41, Q 5, May/June 2018)

11 Mixtures can be separated in various ways depending on the physical properties of their components.
A student is supplied with two different mixtures. The first is a mixture of two solids, sodium chloride and sand. The second is a mixture of two liquids, ethanol and butanol.
For each mixture, describe a method to obtain a pure sample of each substance in the mixture. In your description you should include the names of any techniques and apparatus used.

a solid sodium chloride and sand [5]

b ethanol and butanol [5]

[Total: 10]

(Cambridge O Level Chemistry (5070) Paper 41, Q 6, May/June 2018)

12 A student is given three colourless liquids, **A**, **B** and **C**.

She knows that the liquids are hexene, ethanol and ethanoic acid.

She needs to identify **A**, **B** and **C**.

She does some tests by adding the reagents in the table to **A**, **B** and **C**.

Some of her results are recorded in the table.

test	observations		
	A	**B**	**C**
Add aqueous bromine		The mixture turns from orange to colourless.	The mixture remains orange
Add solid calcium carbonate	No visible change	No visible change	
Add dilute sulfuric acid and a few drops of aqueous potassium manganate(VII).		The mixture turns from purple to colourless.	The mixture remains purple.

a Use the observations in the table to identify liquids **A**, **B** and **C**. [2]

b Complete the table. [3]

c The student mixes two of the three liquids together, adds a few drops of concentrated sulfuric acid as a catalyst and warms the mixture. A sweet smell is produced.

  i Name the two liquids that the student mixes. [1]

  ii Suggest a safety precaution the student should take when doing this experiment. Give a reason for your answer. [1]

[Total: 7]

(Cambridge O Level Chemistry (5070) Paper 41, Q 3, May/June 2019)

Glossary

Acid An acid is defined as a proton donor. Acid substances dissolve in water, producing $H^+(aq)$ ions as the only positive ions.

Activation energy The activation energy, E_a, is the minimum energy that colliding particles must have in order to react.

Addition polymer A polymer formed by an addition reaction. For example, poly(ethene) is formed from ethene.

Addition reaction A reaction in which an atom or group of atoms is added across a carbon-carbon double bond.

Alcohols Organic compounds containing the –OH group. They have the general formula $C_nH_{2n+1}OH$. Ethanol is by far the most important of the alcohols and is often just called 'alcohol'.

Alkali A soluble base which produces $OH^-(aq)$ ions in water.

Alkali metals The six metallic elements in Group I of the Periodic Table.

Alkaline earth metals The six metallic elements in Group II of the Periodic Table.

Alkanes A family of saturated hydrocarbons with the general formula C_nH_{2n+2}.

Alkenes A family of unsaturated hydrocarbons with the general formula C_nH_{2n}.

Allotropy The existence of an element in two or more different forms in the same physical state.

Alloy Generally, a mixture of two or more metals (for example, brass is an alloy of zinc and copper) or of a metal and a non-metal (for example, steel is an alloy of iron and carbon, sometimes with other metals included). They are formed by mixing the molten substances thoroughly. Generally, it is found that alloying produces a metallic substance which has more useful properties than the original pure metals it was made from.

Amino acids These naturally occurring organic compounds possess both an $-NH_2$ group and a $-COOH$ group on adjacent carbon atoms. There are 20 naturally occurring amino acids, of which glycine is the simplest.

Amphoteric oxide An oxide which can behave as an acid (it reacts with an alkali) or a base (it reacts with an acid), for example zinc oxide, ZnO, and aluminium oxide, Al_2O_3.

Anaerobic decay Decay which takes place in the absence of oxygen.

Anaerobic respiration Respiration that takes place in the absence of air.

Anhydrous substance Substance that contains no water. For example, anhydrous salts which have had their water of crystallisation removed by heating them.

Anions Negative ions; these are attracted to the anode.

Anode The positive electrode. It is positively charged because electrons are drawn away from it.

Artificial fertiliser A substance added to soil to increase the amount of elements such as nitrogen, potassium and phosphorus. This enables crops grown in the soil to grow more healthily and to produce higher yields.

Atmosphere (air) The mixture of gases that surrounds the Earth.

Atmospheric pressure The pressure exerted by the atmosphere on a surface due to the weight of the air above it.

Atom The smallest part of an element that can exist as a stable entity. It has a central nucleus containing neutrons and protons surrounded by electrons in shells. One mole of atoms has a mass equal to the relative atomic mass (A_r) in grams.

Atomic mass unit Exactly $\frac{1}{12}$ of the mass of one atom of the most abundant isotope of carbon, carbon-12.

Avogadro's Law Equal volumes of all gases measured under the same conditions of temperature and pressure contain equal numbers of molecules.

Base A base is defined as a proton acceptor. A base neutralises an acid, producing a salt and water as the only products.

Batch process A batch process is one in which each batch goes through one stage of the production process before moving onto the next stage.

Binary compound A substance that contains two elements chemically combined.

Biodegradable This relates to substances which can be broken down in the environment by, for example, by bacteria.

Blast furnace A furnace for smelting iron ores, such as hematite (Fe_2O_3) and magnetite (Fe_3O_4), to produce cast iron.

Boiling point The temperature at which the pressure of the gas created above a liquid equals atmospheric pressure.

Bond energy An amount of energy associated with a particular bond in a molecular element or compound.

Burette A piece of apparatus used to accurately measure liquid volumes before and at the end of the titration.

Calculating moles of compounds

$$\text{number of moles} = \frac{\text{mass of compound}}{\text{molar mass of compound}}$$

Calculating moles of elements

$$\text{number of moles} = \frac{\text{mass of element}}{\text{molar mass of that element}}$$

Calculating moles of gases One mole of any gas occupies 24 dm^3 (litres) at room temperature and pressure (r.t.p.).

$$\text{number of moles of gas} = \frac{\text{volume of the gas (in dm}^3\text{ at r.t.p.)}}{24\text{ dm}^3}$$

Calculating moles of solutions

$$\text{concentration of a solution (in mol/dm}^3\text{)} = \frac{\text{number of moles of solute}}{\text{volume (in dm}^3\text{)}}$$

Carboxylic acids A family of organic compounds containing the functional group –COOH. They have the general formula $C_nH_{2n+1}COOH$. The most important and well known of these acids is ethanoic acid, which is the main constituent in vinegar. Ethanoic acid is produced by the oxidation of ethanol.

Catalyst A substance which alters the rate of a chemical reaction and is unchanged at the end of the reaction. It increases the rate of a chemical reaction by providing an alternative reaction path which has a lower activation energy, E_a.

Catalytic converter A device for converting dangerous exhaust gases from cars into less harmful emissions. For example, carbon monoxide gas is converted to carbon dioxide gas.

Catalytic cracking The decomposition of higher alkanes into alkenes and alkanes of lower relative molecular mass. The process involves passing the larger alkane molecules over a catalyst of aluminium and chromium oxides, heated to 550°C.

Cathode The negative electrode. It is negatively charged because an excess of electrons move towards it.

Cations Positive ions; these are attracted to the cathode.

Chemical change A permanent change in which a new substance is formed.

Chromatogram The record obtained from chromatography.

Chromatography A technique used for the separation of mixtures of dissolved substances.

Combustion A chemical reaction in which a substance reacts rapidly with oxygen with the production of heat and light.

Compound A substance formed by the combination of two or more elements in fixed proportions.

Complete combustion Reaction of a fuel containing carbon with oxygen to produce only carbon dioxide and thermal energy.

Condensation The change of a gas into a liquid. This process is accompanied by the evolution of heat.

Contact process The industrial manufacture of sulfuric acid using the raw materials sulfur and oxygen.

Condensation polymer A polymer formed by a condensation reaction (one in which water is produced during polymerisation). For example, nylon is produced by the condensaion reaction between 1,6-diaminohexane and hexanedioic acid.

Corrosion The process that takes place when metals and alloys are chemically attacked by oxygen, water or any other substances found in their immediate environment.

Covalent bond A chemical bond formed by the sharing of one or more pairs of electrons between two atoms.

Crystal A solid whose particles are arranged in a definite way giving a specific shape.

Crystallisation The process of forming crystals from a liquid.

Delocalised Refers to spreading out of electrons within the metal structure. The electrons are not attached to any one particular ion.

Diatomic molecule A molecule containing two atoms, for example hydrogen, H_2, and oxygen, O_2.

Diffusion The process by which different substances mix as a result of the random motions of their particles.

Displacement reaction A reaction in which a more reactive element displaces a less reactive element from solution.

Displayed formula A diagram that shows how the various atoms are bonded and shows all the bonds in the molecule as individual lines.

Distillate The condensed vapour produced from a mixture of liquids on distillation.

Distillation The process of boiling a liquid and then condensing the vapour produced back into a liquid. It is used to purify liquids and to separate mixtures of liquids.

Ductile A substance (such as metal) that can easily be pulled into a thin wire.

Electrical conductivity A substance shows electrical conductivity if it allows the passage of an electric current through. Poor electrical conductors are also called electrical insulators.

Electrode A point where the electric current enters and leaves the electrolytic cell. An inert electrode is usually made of platinum or carbon and does not react with the electrolyte or the substances produced at the electrodes themselves.

Electrolysis The decomposition of an ionic compound, when molten or in aqueous solution, by the passage of an electric current.

Electrolyte A substance which will carry electric current only when it is molten or dissolved.

Electron A sub-atomic particle that has a negative charge equal in magnitude to that of a proton. Electrons are found in electron shells which surround the nucleus of the atom.

Electronic configuration (structure) A method of describing the arrangement of electrons within the electron shells of an atom.

Electron shells (energy levels) Electrons group together in electron shells which surround the nucleus of the atom. The electron shells can hold differing numbers of electrons.

Electroplating The process of depositing metals from solution in the form of a layer on other surfaces such as metal.

Electrostatic force of attraction A strong force of attraction between opposite charges.

Element A substance which cannot be further divided into simpler substances by chemical methods.

Empirical formula A formula showing the simplest whole number ratio of the different atoms or ions present in a compound.

End-point The end-point of a titration is reached when the acid and base have just neutralised each other. This point is shown when the indicator changes colour suddenly.

Endothermic reaction An endothermic reaction absorbs thermal energy from the surroundings leading to a decrease in the temperature of the surroundings.

Enthalpy Energy stored in chemical bonds, given the symbol H.

Enthalpy change of reaction The transfer of thermal energy during a reaction is called the enthalpy change, ΔH, of the reaction. ΔH is negative for exothermic reactions and positive for endothermic reactions.

Enzymes Protein molecules produced in living cells. They act as biological catalysts and are specific to certain reactions. They operate only within narrow temperature and pH ranges.

Equilibrium A chemical equilibrium is a dynamic state in which the concentrations of the reactants and products remain constant because the rate at which the forward

reaction occurs is the same as that of the back reaction in a closed system.

Esters A family of organic compounds formed by the reaction of an alcohol with a carboxylic acid in the presence of concentrated H_2SO_4.

Eutrophication A process that occurs when fertiliser drains into lakes and rivers, causing algae to multiply rapidly and the water to turn green. it results in fish and other organisms suffocating and dying through lack of oxygen in the water.

Evaporation A process occurring at the surface of a liquid involving the change of state of a liquid into a gas at a temperature below the boiling point. When a solution is heated the solvent evaporates and leaves the solute behind.

Excess reactant A reactant is in excess when, after the reaction is complete, some of it remains unreacted.

Exothermic reaction An exothermic reaction transfers thermal energy to its surroundings leading to an increase in the temperature of the surroundings.

Fermentation A series of biochemical reactions brought about by the enzymes in yeast or, more generally, by micro-organisms.

Filtrate The liquid or solution which passes through a filter during filtration.

Filtration The process of separating a solid from a liquid using a fine filter paper which does not allow the solid to pass through.

Flue gas desulfurisation (FGD) The process by which sulfur dioxide gas is removed from the waste gases of power stations by passing them through calcium hydroxide slurry.

Fossil fuels Fuels, such as coal, oil and natural gas, formed from the remains of plants and animals.

Fraction A mixture of hydrocarbon compounds that boil at similar temperatures and have similar chain lengths.

Fractional distillation A distillation technique used to separate a mixture of liquids that have different boiling points.

Fuel cells In a hydrogen-oxygen fuel cell, hydrogen and oxygen are used to produce a voltage, and water is the only product.

Functional group The atom or group of atoms responsible for the characteristic reactions of a compound. For example, the alcohol functional group is –OH.

Gas syringe A piece of apparatus used to accurately measure the volume of a gas.

Giant covalent structure A substance containing thousands of atoms per molecule.

Giant ionic lattice A lattice of positive and negative ions held together by the electrostatic forces of attraction between ions.

Group A vertical column of the Periodic Table containing elements with the same number of electrons in their outer shells. Elements in a particular group show trends in their properties, both chemical and physical.

Haber process The chemical process by which ammonia, NH_3, is made in very large quantities from nitrogen and hydrogen.

Halogens The elements found in Group VII of the Periodic Table.

Homologous series A series of compounds which possess the same functional group and have the same general formula. The members of the series differ from one member to the next by a $-CH_2-$ unit and show a trend in physical properties as well as having similar chemical properties.

Hydrated substance A substance that is chemically combined with water. For example, hydrated salts which contain water molecules in their crystal structures.

Hydrocarbon Compounds which contains atoms of carbon and hydrogen only.

Immiscible When two liquids form two layers when mixed together, they are said to be immiscible.

Incomplete combustion Combustion that takes place when there is not enough oxygen present for complete combustion. Carbon monoxide is produced instead of carbon dioxide.

Indicator A substance used to show whether a substance is acidic or alkaline (basic), for example thymolphthalein.

Inert electrode These are electrodes that do not react with the electrolyte or the products of electrolysis. Examples of inert electrodes are carbon and platinum.

Insoluble If a solute does not dissolve in a solvent it is said to be insoluble.

Intermolecular bonds These are attractive forces which form between neighbouring molecules.

Ion An atom or group of atoms which has either lost one or more electrons, making it positively charged, or gained one or more electrons, making it negatively charged.

Ionic bond A strong electrostatic force of attraction between oppositely charged ions.

Ionic equation The simplified equation of a reaction which we can write if the chemicals involved are ionic substances.

Ionisation The process whereby an atom gains or loses an electron(s) to become an ion.

Isomers Compounds which have the same molecular formula but different structural arrangements of the atoms.

Isotopes Different atoms of the same element that have the same number of protons but different numbers of neutrons. Isotopes of the same element have the same chemical properties because they have the same number of electrons and therefore the same electronic configuration.

Kinetic particle theory A theory which accounts for the properties of matter in terms of its constituent particles.

Lattice A regular three-dimensional arrangement of atoms/ions in a crystalline solid.

Law of conservation of mass The total mass of reactants created in a reaction is equal to the total mass of product.

Law of constant composition Compounds always have the same elements joined together in the same proportion.

Limiting reactant The reactant that limits the amount of product that can form.

Locating agent A substance used to locate, on a chromatogram, the separated parts of a mixture in chromatography.

Malleable A substance (such as metal) that can easily be bent or hammered into different shapes.

Mass number (nucleon number) Symbol *A*. The total number of protons and neutrons found in the nucleus of an atom.

Matter Anything which occupies space and has a mass.

Melting point The temperature at which a solid begins to turn into a liquid. Pure substances have a sharp melting point.

Metal extraction The method used to extract a metal from its ore; it depends on the position of the metal in the reactivity series.

Metallic bond An electrostatic force of attraction between the 'sea' of electrons and the regular array of positive metal ions within a solid metal.

Metalloid (semi-metal) An element with properties between those of metals and non-metals, for example boron and silicon.

Metals A class of chemical elements which have a characteristic lustrous appearance and are good conductors of heat and electricity.

Miscible When two liquids form a homogeneous layer when mixed together, they are said to be miscible.

Mixture Two or more substances mixed together that can be separated by physical means.

Molar gas volume One mole (1 mol) of a gas occupies 24 dm³ at room temperature and pressure, r.t.p.

Molar mass The mass of 1 mole of a compound, it has units of g/mol.

Mole The amount of substance which contains 6×10^{23} atoms, ions or molecules. This number is called Avogadro's constant.

Molecular formula A formula showing the number and type of different atoms in one molecule.

Molecule A group of atoms covalently bonded together.

Monatomic molecule A molecule which consists of only one atom, for example neon and argon.

Monomer A simple molecule, such as ethene, which can be polymerised.

Neutralisation The process in which the acidity or alkalinity of a substance is destroyed. Destroying acidity means removing $H^+(aq)$ by reaction with a base, carbonate or metal. Destroying alkalinity means removing the $OH^-(aq)$ by reaction with an acid. $H^+(aq) + OH^-(aq) \rightarrow H_2O(l)$.

Neutron An uncharged sub-atomic particle present in the nuclei of atoms.

Noble gases The unreactive gases found in Group 0 of the Periodic Table.

Non-metals A class of chemical elements that are typically poor conductors of heat and electricity.

Non-renewable energy sources Sources of energy, such as fossil fuels, which take millions of years to form and which we are using up at a rapid rate.

Nucleus Found at the centre of the atom, it consists of protons and neutrons.

Oil refining The general process of converting the mixture that is collected as petroleum into separate fractions. The fractions are separated from the petroleum mixture by fractional distillation.

Ore A naturally occurring mineral from which a metal can be extracted.

Organic chemistry The branch of chemistry concerned with compounds of carbon found in living organisms.

Organic compounds Substances whose molecules contain one or more carbon atoms covalently bonded with another element (including hydrogen, nitrogen, oxygen, the halogens as well as phosphorus, silicon and sulfur).

Oxidation Gain of oxygen; can also be defined in terms of loss of electrons or an increase in oxidation number. An element may combine with oxygen to form an oxide.

Alternatively, an atom or ion may be oxidised by losing an electron. In both cases, there is an increase in the oxidation number of the substance that is oxidised.

Oxidation number The oxidation number of an element is the total number of electrons that an atom either gains or loses in order to form a chemical bond with another atom.

Oxidising agent A substance that oxidises another substance and is itself reduced during a redox reaction. It will cause an increase in oxidation number.

Particulates Very small particles found in the atmosphere, such as certain types of smoke emitted from diesel engines, as well as dust.

Percentage composition The proportion of a particular element in a compound by mass, written as a percentage.

Percentage purity The percentage of a specified compound or element in an impure sample.

Percentage yield

$$\text{percentage yield} = \frac{\text{actual yield}}{\text{theoretical yield}} \times 100$$

Periodic Table A table of elements arranged in order of increasing proton number to show the similarities of the chemical elements with related electronic configuration.

Periods Horizontal rows of the Periodic Table. Within a period the atoms of all the elements have the same number of occupied shells but have an increasing number of electrons in the outer shell.

Petroleum Petroleum is a mixture of hydrocarbons. It is often referred to as crude oil.

pH scale A scale running from 0 to 14, used for expressing the acidity or alkalinity of a solution.

Photosynthesis The chemical process by which green plants synthesise their carbon compounds from atmospheric carbon dioxide using light as the energy source. Photosynthesis can only take place when chlorophyll is present.

Physical change A change in a substance that does not involve forming new chemical bonds, so no new substance is formed. The change is easily reversed.

Pipette A piece of apparatus used to accurately measure a fixed volume of a given solution, for example 25 cm³ or 50 cm³

Plastic A man-made material made from polymers.

Pollution The modification of the environment caused by human influence. It often causes harm to living things. Atmospheric pollution is caused by gases such as sulfur dioxide, carbon monoxide and nitrogen oxides being released into the atmosphere by a variety of industries and also by the burning of fossil fuels. Water pollution is caused by many substances, such as those found in fertilisers and in industrial effluent.

Polymer A substance possessing very large molecules consisting of repeated units or monomers. Polymers therefore have a very large relative molecular mass.

Polymerisation The chemical reaction in which molecules (monomers) join together to form a polymer.

Polyamide A condensation polymer, such as nylon, that contains the amide link, –NHOC–.

Proteins Natural polyamides formed from amino acids by condensation reactions.

Proton number (atomic number) The number of protons in the nucleus of an atom. Symbol Z. The proton or atomic number is also equal to the number of electrons present in an atom.

Proton A sub-atomic particle that has a positive charge equal in magnitude to that of an electron. Protons occur in all nuclei.

Pure substances These are substances that are made of only one type of atom, compound or one type of molecule (a group of atoms bonded together).

Purity The measure of whether a substance is pure.

Radioactive A property of unstable isotopes. They disintegrate spontaneously to release energy and one or more types of radiation.

Radioisotope A radioactive isotope.

Reaction pathway diagram A graph that shows how much thermal energy is transferred between reactants and products. The reaction pathway is shown on the x-axis. The difference in energies between the energy of reactants and products on the y-axis represents the overall enthalpy change of the reaction.

Reaction rate A measure of the change which happens during a reaction in a single unit of time. It may be affected by the following factors: surface area of the reactants, concentration of the reactants, the pressure of gases, the temperature at which the reaction is carried out and the use of a catalyst.

Reactivity series An order of reactivity, giving the most reactive metal first, based on results from experiments with oxygen, water and dilute hydrochloric acid.

Redox reaction A reaction which involves simultaneous reduction and oxidation.

Reducing agent A substance which reduces another substance and is itself oxidised during a redox reaction. Also it will cause a decrease in oxidation number.

Reduction The loss of oxygen; can also be defined in terms of gain of electrons or a decrease in oxidation number. A substance that is reduced in a reaction may lose oxygen, for example a metal oxide is reduced to the metal. Alternatively, an atom or ion may be reduced by gaining an electron. In both cases there is an increase in the oxidation number of the substance that is reduced.

Relative atomic mass Symbol A_r.

$$A_r = \frac{\text{average mass of isotope of the element}}{\frac{1}{12} \times \text{mass of 1 atom of carbon-12}}$$

Relative molecular mass The sum of the relative atomic masses of all those elements shown in the formula of the substance. Symbol M_r. For ionic compounds this is referred to as the relative formula mass.

Renewable energy Sources of energy which cannot be used up or which can be made at a rate faster than the rate of use.

Residue The substance that remains after evaporation, distillation or filtration have taken place.

R_f value The ratio of the distance travelled by the solute to the distance travelled by the solvent in chromatography.

$$R_f = \frac{\text{distance travelled by solute}}{\text{distance travelled by solvent}}$$

Reversible reaction A chemical reaction which can go both forwards and backwards. This means that once some of the products have been formed they will undergo a chemical change once more to re-form the reactants. The reaction from left to right in the equation is known as the forward reaction and the reaction from right to left is known as the reverse reaction.

Rust A loose, orange–brown, flaky layer of hydrated iron(III) oxide found on the surface of iron or steel. The conditions necessary for rusting to take place are the presence of oxygen and water. The rusting process is encouraged by other substances such as salt. It is an oxidation process.

Rust prevention To prevent iron rusting, it is necessary to stop oxygen and water coming into contact with it. The methods employed include painting, oiling/greasing, coating with plastic, plating, galvanising and sacrificial protection.

Saturated A saturated compound has molecules in which all carbon–carbon bonds are single bonds.

Saturated solution A solution containing the maximum concentration of a solute dissolved in the solvent at a given temperature.

Simple molecular substance A substance that possesses between one and a few hundred atoms per molecule.

Slag A waste material, mainly calcium silicate, produced in a blast furnace when iron is extracted from iron ore.

Smog Caused by atmospheric pollutants that can react in sunlight with other substances to produce a hazy, harmful atmosphere.

Soluble If a solute dissolves in a solvent it is said to be soluble.

Solute The substance that is dissolved in a solvent.

Solution A liquid mixture composed of two or more substances.

Solubility The solubility of a solute in a solvent at a given temperature is the number of grams of the solute which can dissolve in 100 g of solvent to produce a saturated solution at that temperature.

Solvent A substance that dissolves a solute to form a solution.

$$\text{solute} + \text{solvent} \xrightarrow{\text{dissolves}} \text{solution}$$

State of matter The form that particles of a substance take depending on the arrangement and movement or energy of the particles. The three states of matter are solid, liquid and gas. Changes of state can be explained by changes in the motion and arrangement of particles.

Strong acid An acid which produces a high concentration of $H^+(aq)$ ions in water solution, for example hydrochloric acid.

Structural formula A formula that is an unambiguous description of the way the atoms are arranged, including the functional group.

Substitution reaction A reaction in which an atom or group of atoms is replaced by another atom or group of atoms.

Theoretical yield The maximum possible amount of product which could be formed from the amounts of reactants, as predicted by the balanced equation for the reaction.

Thermal conductivity How well a substance transfers thermal energy. Good thermal conductors transfer thermal energy quickly. Poor thermal conductors are also called thermal insulators.

Titrant is a solution of known concentration that is added (titrated) to another solution to determine the concentration of a second chemical substance.

Titration A method of volumetric analysis in which a volume of one reagent (for example an acid) is added to a known volume of another reagent (for example an alkali) slowly from a burette until an end-point is reached. If an acid and alkali are used, then an indicator is used to show that the end-point has been reached.

Toxic Harmful to health, even deadly.

Transition elements The elements found in the centre of the Periodic Table, between Groups II and III.

Trend A term used by chemists to describe how a particular property changes across a period or down a group of the Periodic Table.

Unsaturated An unsaturated compound has molecules that possess at least one carbon–carbon double or triple covalent bond.

Viscosity How easily a liquid flows. A viscous liquid does not flow easily.

Volatile How easily a liquid evaporates (turns to a gas).

Water of crystallisation The water molecules present in crystals. This water is incorporated into the structure of substances as they crystallise from solution, for example in copper(II) sulfate pentahydrate ($CuSO_4.5H_2O$).

Weak acid An acid that is partially dissociated in aqueous solution, for example ethanoic acid. It is only partially ionised.

Index

The Periodic Table of Elements

Group

I	II	III	IV	V	VI	VII	VIII
							2 He Helium 4

| 1
H
Hydrogen
1 |

3 Li Lithium 7	4 Be Beryllium 9						
11 Na Sodium 23	12 Mg Magnesium 24						
19 K Potassium 39	20 Ca Calcium 40	21 Sc Scandium 45	22 Ti Titanium 48	23 V Vanadium 51	24 Cr Chromium 52	25 Mn Manganese 55	26 Fe Iron 56
27 Co Cobalt 59	28 Ni Nickel 59	29 Cu Copper 64	30 Zn Zinc 65

5
B
Boron
11 — 6
C
Carbon
12 — 7
N
Nitrogen
14 — 8
O
Oxygen
16 — 9
F
Fluorine
19 — 10
Ne
Neon
20

13
Al
Aluminium
27 — 14
Si
Silicon
28 — 15
P
Phosphorus
31 — 16
S
Sulfur
32 — 17
Cl
Chlorine
35 — 18
Ar
Argon
40

37
Rb
Rubidium
85 — 38
Sr
Strontium
88 — 39
Y
Yttrium
89 — 40
Zr
Zirconium
91 — 41
Nb
Niobium
93 — 42
Mo
Molybdenum
96 — 43
Tc
Technetium
101 — 44
Ru
Ruthenium
101 — 45
Rh
Rhodium
103 — 46
Pd
Palladium
106 — 47
Ag
Silver
108 — 48
Cd
Cadmium
112

31
Ga
Gallium
70 — 32
Ge
Germanium
73 — 33
As
Arsenic
75 — 34
Se
Selenium
79 — 35
Br
Bromine
80 — 36
Kr
Krypton
84

55
Cs
Caesium
133 — 56
Ba
Barium
137 — 57–71
lanthanoids — 72
Hf
Hafnium
178 — 73
Ta
Tantalum
181 — 74
W
Tungsten
184 — 75
Re
Rhenium
186 — 76
Os
Osmium
190 — 77
Ir
Iridium
192 — 78
Pt
Platinum
195 — 79
Au
Gold
197 — 80
Hg
Mercury
201

49
In
Indium
115 — 50
Sn
Tin
119 — 51
Sb
Antimony
122 — 52
Te
Tellurium
128 — 53
I
Iodine
127 — 54
Xe
Xenon
131

87
Fr
Francium
226 — 88
Ra
Radium
226 — 89–103
actinoids — 104
Rf
Rutherfordium — 105
Db
Dubnium — 106
Sg
Seaborgium — 107
Bh
Bohrium — 108
Hs
Hassium — 109
Mt
Meitnerium — 110
Ds
Darmstadtium — 111
Rg
Roentgenium — 112
Cn
Copernicium

81
Tl
Thallium
204 — 82
Pb
Lead
207 — 83
Bi
Bismuth
209 — 84
Po
Polonium — 85
At
Astatine — 86
Rn
Radon

113
Nh
Nihonium — 114
Fl
Flerovium — 115
Mc
Moscovium — 116
Lv
Livermorium — 117
Ts
Tennessine — 118
Og
Oganesson

Key

a
X
b

a = proton (atomic) number
X = atomic symbol
Element name
b = relative atomic mass

lanthanoids

| 57
La
Lanthanum
139 | 58
Ce
Cerium
140 | 59
Pr
Praseodymium
141 | 60
Nd
Neodymium
144 | 61
Pm
Promethium
144 | 62
Sm
Samarium
150 | 63
Eu
Europium
152 | 64
Gd
Gadolinium
157 | 65
Tb
Terbium
159 | 66
Dy
Dysprosium
163 | 67
Ho
Holmium
165 | 68
Er
Erbium
167 | 69
Tm
Thulium
169 | 70
Yb
Ytterbium
173 | 71
Lu
Lutetium
175 |

actinoids

| 89
Ac
Actinium | 90
Th
Thorium
232 | 91
Pa
Protactinium
231 | 92
U
Uranium
238 | 93
Np
Neptunium | 94
Pu
Plutonium | 95
Am
Americium | 96
Cm
Curium | 97
Bk
Berkelium | 98
Cf
Californium | 99
Es
Einsteinium | 100
Fm
Fermium | 101
Md
Mendelevium | 102
No
Nobelium | 103
Lr
Lawrencium |

The volume of one mole of any gas is 24 dm³ at room temperature and pressure (r.t.p.).